To Galileo, the Lynx with the glass eye

THE RING
OF TRUTH

Random House
New York

THE RING OF TRUTH

AN INQUIRY INTO HOW WE KNOW WHAT WE KNOW

PHILIP AND PHYLIS MORRISON

Published in the United States by Random House, Inc.,

New York, and simultaneously in Canada by Random House of Canada Limited, Toronto.

Grateful acknowledgment is made to Cambridge University Press for permission to reprint excerpts from *An Autobiography and Other Recollections* by Cecilia Payne-Gaposchkin, edited by Katherine Haramundanis. Copyright © 1984 by Cambridge University Press.

Library of Congress Cataloging-in-Publication Data

Morrison, Philip.
The ring of truth.

Bibliography: p.
1. Science—Methodology. I. Morrison, Phylis,
1927-   . II. Title.
Q175.M8695   1987   502′.8   87-42646
ISBN 0-394-55663-1

Manufactured in the United States of America

Quality Paperback Book Club ® offers recording on compact discs, cassettes and records. For information and catalog write to QPB Recordings, Dept. 902, Camp Hill, PA 17012.

Design: Robert Aulicino

CONTENTS

THE RING OF TRUTH

The picnic on the infield grass that never was. The picnic scene was videotaped in a studio whose wall and floor were painted a particular shade of green; the playoff ball game between the Mets and the Astros was videotaped in Houston in the usual way. We brought the two images together in a television control room; all the green was dropped electronically from the picnic image, its shape was transferred as a black silhouette to make room inside the ball-park scene, and the two pictures were stitched together. We show for one image what had to be performed for a great many.

1

LOOKING

In Introduction

THIS TEXT AND ITS IMAGES SEEK TO PRESENT a simple but somewhat inside look at parts of the natural sciences from the stance of one who asks, "How do we know?" That question can arise out of simple ever-ready skepticism—the position of Missourians, we are told—but it can also flow with hope and pleasure out of a more positive desire to share in the most important of intellectual tasks, at once difficult and delicious, that of gaining and assessing knowledge.

I am a theoretical physicist, fifty years in physics since my student days. Now and then my hand has turned over the years a little to engineering, especially the improvised but fateful nuclear engineering we undertook in World War II. Beginning about thirty years ago the new astronomy all but captured me, along with many another physicist. Teaching at every level from elementary school to graduate studies has given me the chance to watch in fascination how different people begin to face the activities and concerns of any science.

I do not plan mainly to cite the high authorities of science or to dwell on their strange and grand results. That would not be looking inside science at all. Instead I propose to share openly some of the experience that lies behind any result that has persuaded me, and the major links of argument that surround and follow experience. It is then up to you to ask yourself whether you, too, are persuaded, or not persuaded, by what is before you. Only in such a personal way, the evidence at hand, does anyone come to grasp usefully the new results, the theories, the new devices as they arise, and not simply to learn—or be bewildered by—their intriguing names.

The experience of science can be wonderfully wide: as everyday as a pair of eyeglasses or the hot kitchen oven, as unexpected as the enormous meals of the racing cyclist, as beautiful as the worked gold of ancient craftsmen, or as novel as a video rainbow that discloses the motion of a distant galaxy. We will seek meaning in every experience, to draw out some of the main conclusions—and the doubts as well —of today's natural science.

In the printed word there is an obvious gap in time between authors and readers. No one is in any doubt of that, and in print—not so easily in speech, though the remedy is there in questioning—the reader is not even bound to follow

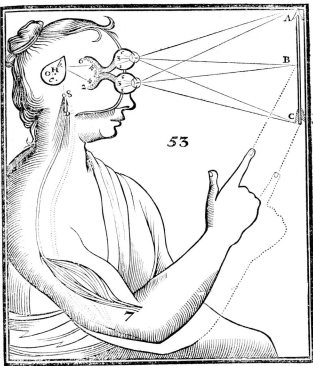

Descartes' eye is the organ of our sense of sight, whose complexities were touched upon in one of his seventeenth-century essays, an early point of view on the senses as inbuilt instruments.

the order of the author's larger design. The coded letters that are strung across the silent page do not in general carry the notion of simultaneity with anything like the force of moving images visible on the screen. Film and television present what looks like the genuine flow of cause and effect, with powerful support from a whole visual context around the center of attention. That richness offers credentials for belief, even though what is there may be, and usually has been, manipulated in ways that range from merely the one-eyed narrow view of the camera to the fully pieced together visual unfolding of a constructed narrative. Words, of course, have similar limitations as evidence, but they do not bring along any external support; they do not much resemble the stream of direct experience we see and hear. Any reader must be more active than the viewer, closer to command over what is going on; that very activity seems to me essential to the gaining of knowledge. It is up to you to decide whether the evidence we present in word and image has that ring of truth.

The mastery held by film and television is in that flow of sewn-together time, obvious in instant replays of TV sports, but present in a great deal of television even when it is by no means evident, even in the nightly news. These powerful channels of communication can ignore cause and effect, or even reverse them. Once the hold of cause is broken, once the unstopping flow of time has been stemmed, it is easy for the filmmaker to piece together into one image the look of different places, and thus to allow us to appear on television to picnic in leisure right among the players of a tense championship ball game that we had never attended at all!

Science is not much concerned with overt deception, as by adroit editors of film and video, whether meant to mislead or simply to entertain. Deception is the fear of a used-car shopper. I am not abandoning the idea of trust. But genuine trust implies the *opportunity* of checking wherever it may be wanted, and trustworthy claims almost always attend to that need one way or another. What science *is* deeply concerned about is something closer to self-deception. It is genuinely hard to make out how the natural world works; any investigator, even a long string of them, can go wrong. That is why it is the evidence, the experience itself and the argument that gives it order, that we need to share with one another, and not just the unsupported final claim.

How We See

IN THIS INITIAL CHAPTER WE TRY TO EXAMINE with some care how it is we all look at the

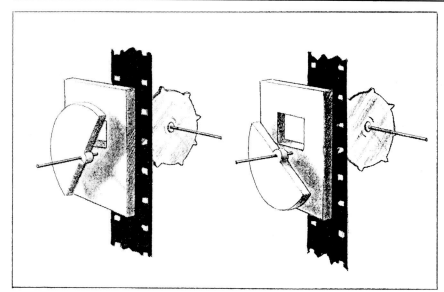

The semicircular shutter of a movie camera blinks closed for half of the time. During the closed period, the film is moved along in readiness for the next image, which is exposed during the open period when the film is held still. A similar intermittent motion is required when the film is projected for viewing. By clever use of filters and colored glasses, or with timed shutters to view through, it would be perfectly possible to show two different movies on the same screen to people sitting side by side.

world. One might expect to distinguish two sorts of looking, one the everyday task we all habitually perform, the other, the specialized close look of the sciences.

But I will not much distinguish between those. I believe those two ways are in great measure parallel, the one quite continuous with the other. Their similarity begins in the fact that both use instruments without which we rarely can judge events. The everyday way of looking uses the human eye, which is after all simply the built-in instrument of human vision. The other way employs a variety of scientific instruments, from the now-familiar telescope to electronic sensors of many sorts, newly contrived every day. A look at how instruments work—both kinds, the inborn and the artificial—and at what they can and cannot do brings a good deal of understanding, whether we ask how we know what happens in daily life or whether we enter the most arcane laboratory of formal science.

What controls knowledge is more the inner nature of the instruments themselves. Deliberate illusions like that of the picnic on the infield, contrived to display what never happened, are not the most important kind.

EVERYONE WOULD AGREE THAT YOU CAN SEE nothing more than a reddened glow of sunlight through the eyelid, and yet we all blink constantly. The time you spend with a closed pair of eyes amounts to 5 percent of all your waking moments. We entirely ignore the loss. For us the flow of vision seems quite unbroken.

It is a little surprising that the ordinary motion picture camera, the work of twentieth-century engineers, blinks out what it might see much more than does the human eye. The image caught on film can certainly be no more than a fraction of what is moving in front of the lens, for the entering light is so much interrupted by an opaque solid metal shutter that the camera necessarily must leave out a third, often as much as a half, of whatever is going on. Yet we piece that sample together in the eye and the mind, without conscious effort, into a fine smooth unbroken stream of moving images. The camera blink is even less troublesome than is the eye blink.

Maybe that suggests a little of how alike the two instruments are, the one fashioned of glass and metal, the other grown of flesh and blood. We always make our judgments of events from an incomplete account, based on assumptions that are somehow built into all our instruments of vision and interpretation. Of course, the motion picture camera was developed, mostly by trial and error, to match the nature of the finally viewing eye.

A photographer can make a flip book of a simple series of snapshots, taken of a sequence of positions, pose after pose, shutter click after

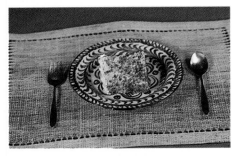

shutter click. When the snapshots are simply presented in order, only much sped up before the eye, the same old flow of motion appears, even though it was not present at all. (Almost everyone has seen a strip of motion picture film that makes plain that every movie is also nothing more than a long string of still pictures, quick regular snapshots quickly unfolded and held one by one before the eye, twenty-four frames each second, the shutter closing between frames to hide the moving blurred image during the time used for replacement.)

It is pretty plain that the eye and the mind want to pick up smooth motion in the world. They have evolved to do so. A staccato, ratchet-like step-by-step shift of position does not often occur in the sort of events that human beings ordinarily encounter. We feel the hefty baseball bat swing smoothly; even if we sample it visually only a few times during the swing, we still insistently build up for it the look that presents the same smooth continuous action we feel. That is the theory of motion present in the eye and mind.

Those moving images are wonderfully powerful. The eye and the mind have nothing to work on but a sample of snapshots with some interruptions. Just the same, they manage to bring us a convincing presentation of continuous motion. Certainly we inherited this ability, and perhaps we train ourselves in its use as well. It seems fair to say, from what we know of

the nature of vision, that we cannot literally *see* smooth motion. Rather we infer that smooth motion is there in the scene, even though the evidence is only those staccato samples, a plain example of the fundamental similarity between everyday experience and the foundations of science: theory extended from experience.

Our everyday, commonsense perception— we recall its subtlety, much more than is in the simple description we have given, and admit also that the system is by no means fully understood—certainly proceeds by the inbuilt instruments we call eye, ear, indeed all the senses. Now compare what the sciences do. There also it is instruments that bring us view after view of various portions of the world around. Scientific instruments are not structures inborn into almost all humans, but devices fashioned over the years by human hands. Often they enable a quite new view that extends and augments those inbuilt senses that evolved biologically, without conscious human aid. The task of judgment that remains, in ordinary affairs as for science, is to fit those many instrumental views together. To pay attention to our ordinary way of looking is at the same time to pay attention to the ways of science, ways that are usually less difficult to fathom than how we all quickly come to know who has entered the room.

Everyday looking is also never a simple task, even though we usually do it quite well. All the

An artful magician can make a crumpled paper ball suddenly disappear—but only to the one person he faces, while every onlooker can easily follow the ball of paper throughout. This skillfully performed sleight-of-hand, so transparent as to seem impossible, depends upon the magician's understanding of how the eye works.

A white square appears here overlapping the four black circles. You will see it if you wait a moment, though it is illusory. Will the effect "work" if the circles are reduced to Vs? How far apart can the circles be? It certainly works for curved contours and contrasts other than black-on-white. Can there be an internal process which infers that what you see is probably another surface, obscuring part of each circle?

same it is instructive to see what happens to our practiced ability to look closely whenever a talented professional appears, one who knows our usual means of looking, but is out to intrigue us by frustrating those familiar expectations we assemble from the sample of the world that we see.

SUCH A PERSON IS THE ENGAGING JEFF McBride, innovative and reflective young mime-magician from New York City. Listen to the conversation we two had not long ago.

MAGICIAN: That's how you can fool our patient subject. The whole audience knows how the trick is done—except for her. For people to understand, not how tricks work, but how their eyes perceive, is the point. If a good magician does an illusion for you, he's going to get you; that's his job. Whereas your job as a scientist is to uncover things, to reveal things, we try to take all of science, but hide as much of it as we can. We use everything that science teaches against our audience.

PHYSICIST: But you see that those are two sides of a coin.

MAGICIAN: Yes, the same coin.

PHYSICIST: The same coin, the coin of perception, whose other side is misdirection, which nature produces even when you, the clever magician, do not.

MAGICIAN: We try to reproduce events that cannot possibly happen in nature, using the information drawn from nature through the scientist.

PHYSICIST: Yes, you deal both in causes that are not allowed to have effects and in effects whose causes are concealed. When you put them together, the onlookers think the cause produces the effect even when palpably it cannot. The rabbit is not in the hat.

MAGICIAN: The rabbit is not in the hat. You use people's information against them. People have only two eyes, and they move together, only in a single direction at a time. They can't focus on two things at once. What I do in this particular effect is seduce their eyes with a slow motion by one hand, while the other hand is moving very quickly. When you are looking here into my hand you can't see the ball of paper go out of play right over your head.

But then again, you've been watching this paper ball go up and down all morning; whenever it was tossed, it just plain vanished from sight. Now you know how that happens, so what we have to do is recognize that and exploit your preconception of how that paper acts. Then I will change it, so what you expected to vanish quietly became all of a sudden a startlingly visible transformation!

Here the Magician indeed fooled and star-

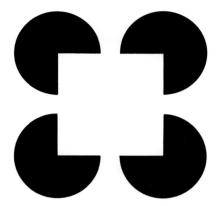

tled the ingenuous physicist, exactly as he had promised. Instead of the tossed ball of paper, moving away from me as gently as it had arched before the onlookers time after time, there was a contrasting event that I had never expected. A burst of white confetti streamed out from my startled gaze. The substitution was spectacular, if obvious enough after the fact.

The magician had exploited one more piece of his insight. How could his confetti fly out so swiftly from a light toss of the hand? That was not a simple tangle of paper ribbon; the paper tape turned out to be loaded at many points with tiny lead shot, whose weight very much speeded the motion of the white streamers through the air.

YOU ALWAYS LOOK STRAIGHT OUT TOWARD THE wall ahead of you to see your own face; a mirror there has bounced the light back. If it were not so commonplace, it would seem a paradox to see by looking away. A mirror of glass put frankly into the light path can certainly influence the visual judgment, but there it is, external, visible, tangible, and somehow easy to understand.

Visual judgments require internal processing in the eye and the brain. Sometimes that is conscious; sometimes it is a process subjectively quite unknown to us, but it is pretty surely present. Some effects found only during the last decade or two make that evident. The

drawing shows four pielike black disks, well arranged on a completely fresh and blank sheet of paper. Look closely for a few seconds at the group of disks to allow time for internal processing. A second sheet faintly lighter than the background seems to cover the segments of the four disks, and the lighter superimposed sheet can be made out at its boundary as well. That appearance is internal. It is not in the paper; it is not in the disks. It is inside the viewer.

We can change the variables. Try red instead of black contrast disks; the cuts are curved instead of straight; three disks form a triangle, instead of four to make a square. Once again you need to allow a little time for the internal processing that is unconsciously carried out in the eye and brain. The superimposed screen appears again, perhaps with less contrast than with the black disks, but now with a curved edge contour, no longer a simple straight line. A remarkable kind of processing is going on.

These illusions, discovered by Gaetano Kanizsa at Turin, have been studied well by now. What has been learned can be explained by the following simple proposal. The visual system inside the eye and the brain somehow intends to describe what we see in these patterns by an act of unification, imposing a single screen above the cut disks, as if it were a second sheet of paper appearing a little whiter, a little different from the background. In a way the scene is carried into the third dimension.

SEEING DEPTH WHERE NONE EXISTS

AN IMPRESSION OF DEPTH, OF THE third dimension, is built up for nearby objects by the way we use our two eyes. The distance between the eyes allows us to see slightly different aspects of the same nearby scene; inner computations on those paired images give us strongly our sense of the distance of objects, and of their three-dimensional shape. At greater distances, other clues take over, such as making out detail and the atmospheric graying of distant objects. We are not born with all that skill: as infants we must experiment, reach for the crescent moon, try, fail, and slowly come to know the shape of the world, especially at some distance. The nearby stereo judgment seems inborn.

But the pair of images presented here make a strong challenge to such a system. If you look at them with a hand lens, you will see they are made up of many little triangles, black and white. They are arrayed identically in the two images, except that in the center of the right pattern a whole triangular array is shifted somewhat to the left. As you look at the images one at a time, it is difficult to tell where the shift is, where its edges are. But if you look with a stereoscope, or fuse the two images into one by the method described here, somehow you sort out without conscious effort the complete pair of random patterns, and see a large speckled triangle floating above a speckled backdrop. Many minute differences and similarities of the whole array have to be processed to do that. But it takes a little time before that image appears, a sure sign that an internal process had to work its way through and correlate the positions of all those speckles.

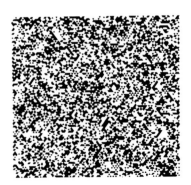 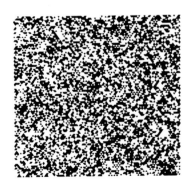

A random-pattern stereo pair. It is sometimes possible to see a stereoscopic effect without an instrument made for that purpose. Pretend to look *through* the two pictures, focusing distantly. Keep each eye from seeing the other's picture by holding a sheet of paper between your nose and the book. Have bright, bland light fall on the images.

A personal corner-box can be made by copying this pattern on paper, then cutting, folding and taping it into shape. Cradle it, corner pointing away from you, in the palm of your hand. Hold it at arm's length and close one eye. After a moment or two the inside corner-box should appear to be the protruding corner of a solid cube; then move it slowly from side to side: your eye will see one motion, although your hand will know another. The exhibit of many fixed corner-boxes looks the same; it is lighted from below.

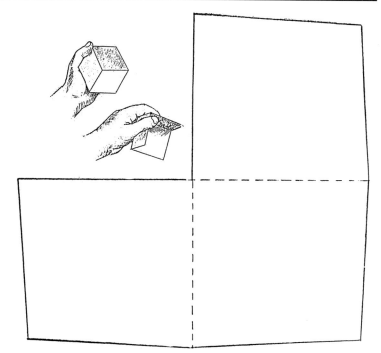

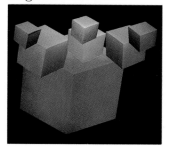

To cover one hand with the other requires a little thickness, if not very much. The superimposed screen we add here, like the two hands one on the other, extends a little out of the flat two-dimensional surface of disks and paper. That third dimension—only a little of it—is what we appear to add on our own in the unconscious internal processing we all carry out upon such images.

THE SET OF BOXES PROVIDES A STILL MORE overt test of our perception of three-dimensional space. Most people will see the boxes in the photograph as lighted from above. In fact they are lighted from below; a hand from below can cast a shadow, from above it will not. These are not the outsides of small boxes, they are hollows, the interior corners, the insides of cubical boxes. When I confronted the real three-dimensional set of boxes with one eye closed—as the camera is one-eyed—I saw the box tops lighted, the corners protruding toward the viewer, a dozen ordinary cubical boxes arranged for one to see. When I moved my head the falsely seen boxes moved quite strangely, as if they followed my own motion. I had simply inverted the true nature of their bright little geometrical world.

It is easy to make a simpler version, just as striking. All you need is a part of a cubical box made of paper or cardboard (the pattern is on the page), only three of the six cube faces, held in the hand to present the interior corner. If I hold the corner in my hand in almost any lighting I can easily persuade my open eye (one is closed)—in fact I can hardly do otherwise—that I am looking not into an interior corner, but at an outside corner of a cube. The conviction is so strong that when I move the piece slightly before my eye its motion seems entirely contrary, as though it were an animate thing in my hand. My eye and brain have inverted the facts of 3-D space, and with that inverted the appearance of motion. (The reader ought not to miss the riveting experience of building the little cube to try it out.)

What has happened is in part quite plain. The two-dimensional retina cannot gain enough geometrical information from the simple bland white cube to decide which way that corner faces, inward or outward. It has made the wrong three-dimensional inference. We always try to judge a three-dimensional world from inadequate cues; sometimes we judge wrongly.

PERSPECTIVE

AN INFANT REACHES EAGERLY FOR the bright moon, "very like a sixpence." The child does not yet grasp perspective, we might say. The image of the moon we catch in the eye is about the same as that of a coin or a ball that is well within reach. Experience teaches us that the moon is far beyond our grasp (save by three-day rocket trip), but a little person who cannot walk about does not know that.

The nub of the idea of perspective is the perception of three-dimensional space by the use of a mere surface-thin image that falls on a screen behind a lens, whether it falls on a human retina or on the film in the camera. It is a rather abstract notion, as the history of painting makes clear. The painters of ancient and medieval times, and to a degree Asian artists even much later, did not employ this particular abstraction. It arises from the cone of rays along which the light from a distant object comes to the watching eye. By that piece of geometry the two rails half a mile down the railroad track are set very close indeed, while those at our feet are comfortably wide-spaced.

The camera agrees with the eye, of course, and so did the clever painters of the Renaissance who first of all codified the rules by which such a structure of converging lines is made pretty faithful to the scene. Some such rules are built into the lens-retina system as well, for we see the same converging track the camera shows. But you can be sure that no Amtrak locals ever rattled down such a narrowing track.

That is only one way of adding a three-dimensional meaning to the eye's grasp. The eye is fooled by the moon, an object so far and large as to lie outside ordinary rules. The eye is fooled when it examines a piece of architecture—built first by the waggish architects of the Baroque, and later even more carefully by Adelbert Ames at Dartmouth—which avoids equal-length columns, horizontal ceilings, and rectangular windows, but introduces instead all sorts of slants and distortions to generate a projected image that fits exactly the message the light would bring from a normal set of solid shapes.

There are many remedies for this misper-

The rules of perspective drawing were soon mastered by the artists. Here are examples by two adepts of the art, the Flemish de Vries and the Nuremberg master Albrecht Dürer. The method for getting a highly foreshortened image shown in Dürer's woodcut is one he invented himself.

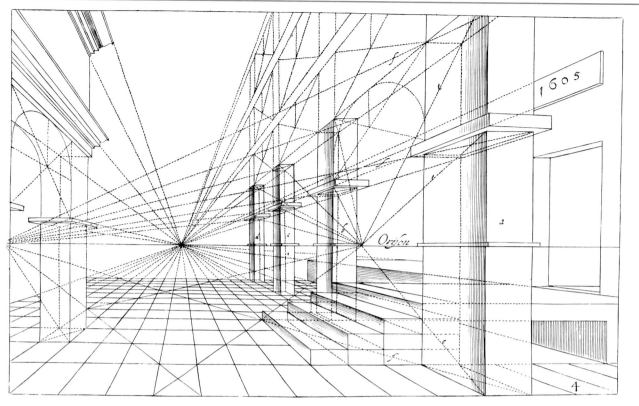

ception. Some of them are built into our own sensory system: the eye lens tries to focus according to distance, the two eyes see slightly different views, and the computer of retina and brain contrive their 3-D stereoscopic explanation. Nearby objects tend to obscure the sight of more distant ones. Above all, the viewer is not constrained to act like an immobilized Cyclops but can take up several points of view spaced wider than any pair of eyes. Even the moon then comes out from behind the hill; it is certainly farther away.

The so-called linear perspective, an approximation to what we inherited with our eyes, seems to have become first consciously understood in the city of Florence soon after 1400 A.D. within the circle of the friends and acquaintances of Filippo Brunelleschi, the architect who designed the dome of the great cathedral there. For a couple of hundred years the single view dominated the painters of Europe, and in our photographic world that style is even more familiar, although now we recognize with Duchamp and Picasso that there are other powerful ways to draw.

Indeed, the old Florentines knew that very well. Perspective is the means used to content the eye by treating the picture surface as a kind of window. But there are other means to represent the world, from the ingenuous way the old artists painted Pharaoh very large among his quite small peasants, to the crisp and cold multiple drawings, scaled like little maps, of the architect and the engineer.

The painter can happily lay out a tiled floor with big tiles nearby and small distorted ones fading into the distance. But the architect, the builders, and the client who foots the bill know very well that the distant tiles are made exactly like the near ones, all of them equal squares. The world we live in is solid, three-dimensional; what we see and photograph is not solid, but literally surficial, and the visual perception of depth within it is governed by subtle rules approximated long ago in painter's perspective. We best reconcile the two worlds by action, by moving about; a change in point of view is a real help to understanding, even in matters far beyond visual representation.

A distorted room made to the specifications of Adelbert Ames in the Science Museum of Virginia. This room looks perfectly rectangular, all as it should be, as long as you view it with one eye from a carefully placed peephole. But if two people enter the room and go to opposite corners, you see an astonishing sight! The room continues to deceive you even when they exchange places. It is not easy to make out what the shape of the room really is, but the final picture shows something nearer to what is really there.

THE STUDY OF HUMAN PERCEPTION IS ONE OF the deepest of sciences, yet in it every human being is an informed participant. The few cases we examined brought surprising results, yet very sensible ones. Take the perception of motion: we do not accept motion as a staccato occupation of point after point—even when that is all the eye sees. On the contrary, we find smooth, coherent changes in the position of moving solid objects. That is the way the world is usually built; the mind and the senses together try to infer a consistent world from our partial perceptions. In space, too, we have similar results. We will not accept any simple projection into the two dimensions of the retina as what we see. No; instead we try to construct a three-dimensional world the best way we can, to infer the world of depth that we expect from what partial information the shallow retina of the eye can bring us.

What we perceive is more than the sense organs can directly tell; it is limited by their structures, it includes the assumptions and the experiences of our kind, both inherited and individual. With scientific instruments as well, such interpretation is almost always a requisite. For science as for everyday, the world is to be inferred on the basis of the signals the instruments get. What we learn is rarely given as a free gift; some effort of our own must enter the transaction.

The determination of the eye-brain to choose a plausible 3-D fit to what is only a 2-D image, or to generate continuous smooth motion out of a sequence of snaps, is rather newly recognized. That recognition belongs to our own age, the time of the computer. Some computerlike features of the human nervous system are at work; we do not know quite what. We know that the process is organic, subtle, intricate.

The most widely used of optical instruments—beyond the looking glass—is a pair of reading glasses. Their task is simple enough, in no way organic or intricate. All the same they arose out of an early interaction between the inbuilt organs of human vision and the growing skills of craftsmen. They augment human vision at the simplest of levels. They allow a sharper focus by an aging natural eye lens by adding another external lens of glass.

Spectacles were first made seven centuries ago, surely from the chance experience of some keenly observant artisan, and not out of any new understanding of the nature of the eye. What is more remarkable is that the very same careful form, the same unusual material, and the same elegant skills that make spectacles commonplace around the world gave rise to an extraordinary landmark among scientific instruments. That was no helpful everyday device at all, but an instrument of rich discovery that rapidly transformed our whole conception of the universe in which we live. What the spectacle makers finally found was the telescope.

Bookmen at work in the library at Treviso, not far from Venice. These frescoes were painted by Tomaso of Modena in the 1350s. Tomaso was in the process of inventing true portraiture, and we can see that art coming into being. Each figure here has an individual task: proofreading, cutting and piecing text, copying, reading, checking details. The artist has represented a famous scholar of a century before using an aid appropriate to an elderly, bookish person, the first pair of glasses whose pictured image has come down to us.

How to See Better

THE TELESCOPE WAS FIRST MADE INTO A POWerful scientific instrument by Galileo Galilei, working in Padua, Venice, and Florence. By its use he first perceived and disclosed within five years a new cosmos, never before seen, that has provided a powerful impetus to all of modern science. His is the universe in which we still find ourselves. But those high matters all began with the homely spectacles.

It is in a library that we would expect to find bookish men using eyeglasses. The paintings shown are in the chapter house of the order that maintained in a small city close to Venice one of the finest manuscript libraries of the fourteenth century. The beauty of these frescoes would have been reason enough to come here. But we came to look at one particular painting that shows a pair of eyeglasses; it is the oldest visual representation of glasses known anywhere. It provided for me irrefutable evidence that the old artist knew and had seen

men who used such glasses. That does not mean that the venerable Dominican whom he paints ever used such glasses; that man was long dead. That is not the kind of evidence you can rely upon from any painting. For the artist had never seen the famous old Latinist; it is unlikely that he had even seen any representation of that man. No, here it is the painter who must have seen eyeglasses, who recognized moreover that they were proper and useful for distinguished scholars, and who painted them in use among the workers in the great library he celebrated here.

These paintings were done somewhat more than six hundred years ago. While that is surely not the earliest use of eyeglasses—the painter would not have arrived in the very place just at the time they were invented; most improbable—still, the first use must have been not so very long before and not so very far away, so that this painting could be the oldest representation of eyeglasses preserved anywhere.

The historians have found evidence for that, too. It is not visual but written evidence. It comes from a sermon that was written down more or less as it was delivered, nearly fifty years before the time of the painting. In Florence, a city some distance away, a visiting priest uttered these words: "It is not twenty years since there was discovered the art of making spectacles to see better, one of the most useful and necessary of arts....I have myself seen and spoken to the man who first discovered and made them." So spoke a priest from Pisa one Sunday morning in the winter of the year 1306.

For me the words of the text are clear enough. They are perhaps not quite as firm as the painting; we don't know just what those first spectacles were like. Were they really so close to the modern form as those we see in the painting? Taking all the evidence together, there is little doubt that the making of "spectacles to see better" began just about 700 years ago somewhere in Tuscany. The same trade, now a considerable industry, still provides this most used of optical instruments all over the world.

Bird's-eye view of Venice in 1500. It is the earliest true map we have of any city. Modern Venice is perfectly recognizable, although this view is unrealistically inactive. This large woodcut map printed from six blocks is an oblique view, seen not from a helicopter but from within the mind of the artist, Jacopo de'Barbari, himself become a bird in his own imagination. Near the center of the map is the broad Square of San Marco with the grandly tall bell tower where Galileo climbed for the vista it gave to his telescope; the island in the Lagoon farthest to the top of the map is Murano, the place of the glass workers then and now.

OUR STORY BEGAN IN THE TIME WHEN THE northern Italian cities were stirring as the curtain fell on the medieval world; the urban trades grew apace, and glassmaking was one of them. It was among the glass workers that spectacles had been invented.

The main act of this long history takes place a full three centuries later, around 1600, in a turbulent time, a time of transition, the time of conscious reaction by the Catholic princes of Europe and the Church of Rome to the shock of the Reformation. The states that faced the Atlantic were beginning to become the chief centers of Europe; Paris and London were growing larger than Milan or Venice or Florence, though Papal Rome still kept pace.

Our time is the time of Shakespeare and Drake, of Rembrandt van Rijn, of Velasquez and of Poussin. No longer could Venice claim to hold "the gorgeous east in fee." But that sea-wedded city-state was still wealthy and powerful; her galleys had led the naval victory that set the limit to the eastward advance of the Turks while our hero, Galileo, was still a child. Venice's musicians at this time included Monteverdi, and her painters included Caravaggio. Florence no longer boasted a Leonardo or a Michelangelo, but the wealthy Medici Grand Dukes continued to rule all Tuscany from their offices in Florence, and the work of Florentine artisans was still unmatched.

The principal actor of the drama is Galileo Galilei, born in nearby Pisa in 1564 into a talented old Florentine family. He became a young professor of mathematics at the University of Pisa, where his unconventional views did not endear him to his colleagues. When his father died, Galileo left Pisa for the distinguished old University of Padua, leaving his native Tuscan state for the Venetian one.

He came to Padua to fill the chair of mathematics at the age of twenty-eight, redheaded, brilliant, energetic, remarkably gifted with his hands, a superb writer, a musician of talent, a good judge of wines, yet a man rather hedged in by family responsibilities. He was to spend nearly twenty years at this celebrated university, the principal center of learning of the

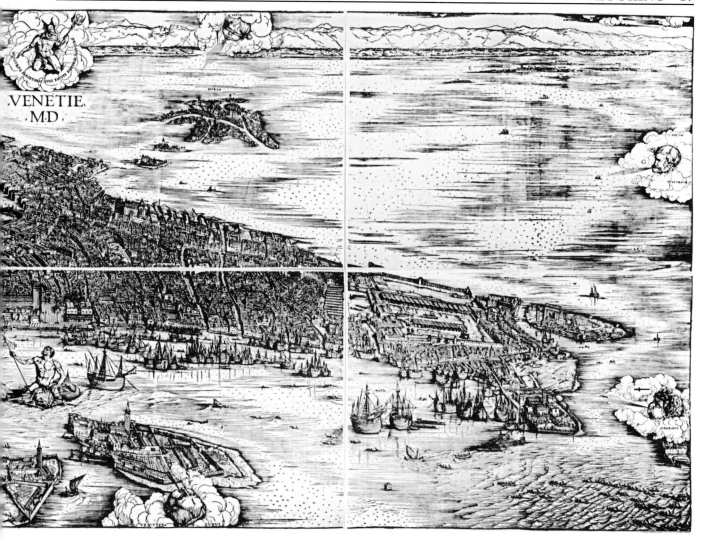

Republic of Venice, constantly complaining about the salary, and not much interested in the required lectures that he gave routinely year after year.

What excited him more was the intellectual life at Padua. From all over Europe there flowed bright students and fine scholars with new ideas and comment on whatever was happening in the life of the mind. William Harvey, for instance, the famous English physician who later discovered the circulation of the blood, took his medical degree at Padua while Galileo was a professor. Somehow Galileo managed to find the means to carry out genuine research and to build the equipment for it—diverse physical apparatus, some of it of a new kind. After a while he was even able to support a skilled craftsman to work along with him. Together they made and sold widely in Europe a very clever calculating device in engraved brass to Galileo's own design. It was to this pair in their precision shop that word came from Venice, the bustling capital city twenty-five miles away, that something new and remarkable had been exhibited there in the early summer of 1609.

That novelty was the telescope.

FOR MORE THAN FOUR HUNDRED YEARS THE island of Murano remained the leading glass-making center in the world. We cannot say for sure, but it is likely that much of the glass that Galileo Galilei ground for his lenses was made in Murano, though some of it was certainly

GLASS

THE TERM *GLASS*, LIKE THE WORD *alloy* or *ceramic*, really denotes a class of materials, rather than any single well-defined substance. There is even natural glass, molten volcanic outflow, like obsidian. But the first appearance of the smooth and lustrous stuff in the furnaces of the artisan was as a glaze, covering or decorating ceramic cores. Glazes go back to earliest times in Egypt, 3,000 or 4,000 years B.C. Glass was used for a very long time as a kind of artificial gemstone, semiprecious material mostly cut and polished like a stone, or as a liquid to build up ceramic or sand cores in a much thicker version of glazing.

Blown glass first appeared in Syria at about the time of Christ. The craft spread with remarkable speed throughout the Roman world, as far as Britain and Russia within a century or two. Its raw materials were sand, limestone, wood ash or another source of alkali, and plenty of fuel, in that day mainly wood charcoal. The forested provinces of Europe were fine places for rural glassmaking. The Romans used window glass, though it was uncommon. By the time of Rome's decline there were glass windows in many a church. Large windows for churches and mansions date to late medieval times.

North Italy, Venice in particular, was a center of artisanship and wealth by the tenth century or so. Venice had moved its glass furnaces before 1200 A.D. to the small neighboring island of Murano, either to abate the nuisance of smoke and fire or to keep the trade secrets of its craftsmen from casual inspection. Murano led European glassmaking for five hundred years and more. More than

Old glass beads, probably from Murano.

A furnace for making and blowing glass, probably the way they did it about 1550 at Murano, with telling detail down to a broken vessel in the foreground.

once some of its workers were coaxed and bribed to move abroad to establish their trade in other lands, including Holland, France, and Bohemia.

Transparent glass cannot be made on too small a scale. The furnace must stay hot a long time, which demands a reserve of fuel; the crucible must be large enough so that the matter from its walls does not contaminate the entire melt; the melt must be held fluid long enough so that bubbles and impurities go off; all the raw materials must be refined and controlled. White sand and carefully cleaned lime and soda ash can make clear glass; even a tiny iron impurity turns the glass greenish.

All these conditions were satisfied within the busy trade in glass vessels and window glass among the cities of north Italy about 1100 or 1200 A.D. About 1280 some clever fellow around Pisa or Florence first recognized the possibility of spectacles. I suspect—there is no proof—that the first glimmer of specs required neither grinding nor polishing: the surface of a chance fragment of clear blown glass was smooth enough and rounded just right so that some observant old glass worker noticed that it clarified his failing vision.

The rest was a cunning art. It is possible that hand-held single magnifying lenses of polished glass or even of gemstone were of much older—and rarer—use among lapidaries and other craftsmen for close-up work. But the use of inexpensive glass lenses, ground to work at a distance and big enough to frame in front of elderly eyes, was the big new idea of 1280. Among churchmen the reading of texts, among merchants the exchange of accounts and letters, and the widespread rise of fine craftsmanship offered ready markets for spectacles; all those professionals wanted to preserve their aging eyesight. Printing and hence reading grew apace in the West after 1450. Spectacles had traveled all the way to the court of China by 1400.

Florentine. He bought some lenses from Venetian craftsmen. Decorative glass is still made on the island by hand in the old craftsmen's way, but not glass for lenses. The fuel is no longer charcoal, but natural gas or oil; the forests nearby are long since gone.

Glass is to this day the essential material of fine lenses. But a disk of clear glass, the product of the glassmakers' furnace, is only the raw material for a lens. The craftsmen who make mirrors or lenses are quite different from those who produce the glass. Glassblowing and molding mean glass worked hot; those operations may best be done next to the furnace. But the patient working of cold glass by grinding, polishing, and cutting is almost never carried out near the fires. How were lenses made before the automatic machines of our time?

We found a man who enjoys making optical lenses by hand in a manner very like that of the spectacle makers of Galileo's day, the path Galileo himself followed when he began in the summer of 1609. Don Dilworth of Massachusetts and Maine is an old-fashioned lens grinder by avocation, but professionally he is a highly skilled designer of the most modern lenses. His complex lens designs and his advanced digital computational schemes are in current use worldwide.

Dilworth uses a small electric motor to do the muscular work of a young apprentice; otherwise, only his materials differ much from the

THE ART OF THE LENS

Lens designer Don Dilworth grinds lenses today almost in the old way.

A lens-grinding lathe used in Florence after the 1620s. Don Dilworth powers his lens grinding with a small electric motor; this one invites driving by 'prentice-power.

From the rough-ground surfaces to the fully polished lens, with one intermediate step on the way.

usages of the seventeenth century. Listen to his account; the procedure we can watch today in Dilworth's basement is what made possible our knowledge of the universe.

The grinding and polishing machine is a tabletop device, mostly of wood. The geared motor steadily turns the mount which receives the lower piece of glass to be worked. The upper piece is held on an arm that can swing back and forth across the lower turning mount in the hand of the artful lens grinder, who places it and presses it against the lower piece as he chooses. The upper glass is free to turn on a bearing of its own, but it is driven only by contact with the piece turning below it. The scheme is only a way of generating portions of a spherical surface.

DILWORTH: To make a lens the process is really very simple. You start off with two disks of glass of approximately the same size. Put the grinding material, dampened abrasive grits, between the two disks. As the machine turns, they grind each other away. Grind with a fairly coarse abrasive until you generate a curved surface with about the right radius.

Clean everything up; switch to a finer grade of grit and grind again for a few minutes. Clean up again, and put even a finer grade of abrasive on the glass. You might finish the grinding only after four or five stages with finer and finer grits.

To grind a concave surface, a glass thinner at the middle and thicker at the edges, you need only to put it in the top mount. Then as you wiggle the swinging arm to and fro, the edge of the top piece will overhang well beyond the turning disk below. There the force is all on a small area, the pressure is higher, and the edge of the bottom piece will grind away effectively at the center of the top piece. You don't spend much time with the top piece centered. For there the pressure is least and the relative motion between the surfaces, which is what does all the work, is small.

DILWORTH: If I want the surface to curve the other way, edges thinner than the center, I would put the blank on the bottom. Whichever surface is on top becomes concave. You control the curve by how you stroke the arm, how much pressure you apply, and what size grits you use.

Once you have worked patiently enough, you have ground a surface with the right curve. But it is not yet a useful lens surface, for it is visibly not at all transparent. It has the translucent matte surface we call ground glass. When you finish the grinding you must begin the second stage, polishing. This step uses not a hard glass tool like the blank itself, but a soft working tool made of black pitch, molded in place to conform to the properly ground surface.

Polishing is done with a much gentler abrasive powder.

DILWORTH: Instead of hard grit particles that roll and grind between two hard glass surfaces, the little particles of polishing compound stick in the soft pitch. They act like little knives that plane the glass to smooth it. This is the most tedious part of the process, but it produces the final glossy surface of the lens.

The quality of my lenses, in spite of this simple equipment, is enormously better than anything Galileo could have made. We have better equipment, much better glass, and better grits. I think it's safe to say that the worst lens that I have ever made down here is far better than anything Galileo could have had.

IT WAS IN THE GREAT PUBLIC SQUARE OF VENICE, the Piazza San Marco, seat of the aristocratic Republic, that Galileo took the first public steps that would lead to the new heavens. To that memorable and beautiful place Albert van Helden, professor of the history of science at Rice University and the author of the best recent study of the development of the telescope, came to talk with me about those events of long ago.

PHYSICIST: So the spectacle-makers did it at last.

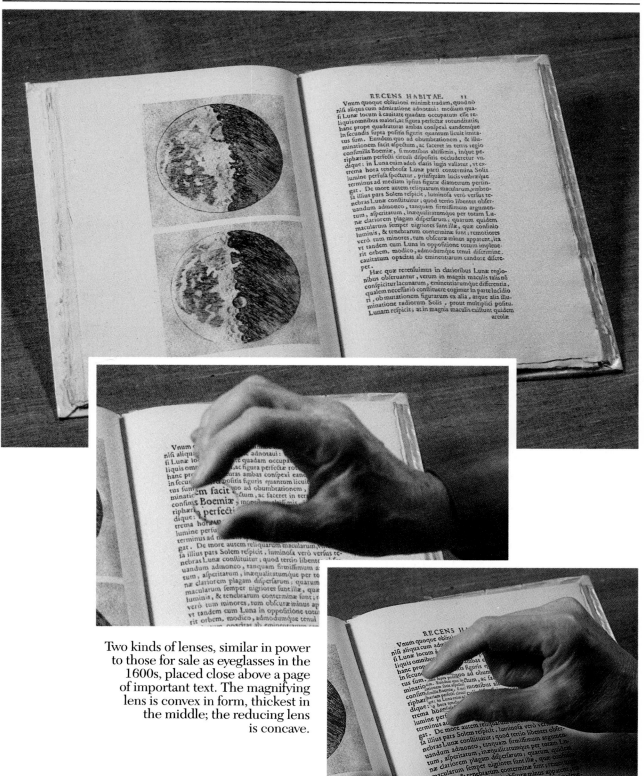

Two kinds of lenses, similar in power to those for sale as eyeglasses in the 1600s, placed close above a page of important text. The magnifying lens is convex in form, thickest in the middle; the reducing lens is concave.

The first telescope. In the San Marco square Van Helden holds the right lenses (the same two shown above the page of text) in the right order at the right distance, to reproduce the optical qualities of the Dutch telescope that came to Venice in 1609.

HISTORIAN: They finally did it. It took them three centuries. Two kinds of lenses—I will call them glasses—are involved. The first one is a magnifying glass, and it dates from around 1300 A.D. It was used by aging scholars whose eyesight was getting steadily weak, as they said then. They used it to enlarge print… no, they didn't have print yet…but to read their written manuscripts.

PHYSICIST: We saw the first painting of a scholar wearing just such glasses.

HISTORIAN: That kind of a glass you would find early amongst the scholarly community. When you hold the glass above the page you can see how it enlarges the print. If you put it close to your eye you can wear it as what we call spectacles. That was the original spectacles, those that use the oldest of lenses, convex, or thickest in the middle.

There is another kind of lens, used by young people—it still is used today—who need to see better at a distance. These are for people who have weak vision of the young, as they used to call it. This glass appears to make things smaller; if you hold it above the page, you see that the letters appear smaller. It is a concave lens, hollowed in the middle, thickest around its edges.

The two together make up the telescope of Galileo. Since both kinds of lenses were around by the middle of the fifteenth century, the question is, if the components were there, why was the instrument not invented? This problem is made more difficult by the fact that the actual telescopic effect is not that difficult to discover.

PHYSICIST: What do you have to do to get the telescope?

HISTORIAN: The effect is very easy to find. If you take the concave glass and hold it up to your eye, to look through it at a distant object like a church steeple, you see it clearly in focus, but smaller. If you then put the convex glass close in front of the other one, you still see things in focus but about the same size. But as you move the convex glass farther and farther away you see a sort of zoom effect: the distant object appears to become larger and larger, until at a certain point the image just dissolves. That is as much magnification as you are going to get; that's the telescopic effect. If I simply hold one of the lenses in each hand, the convex one in front of the other, in effect I have in my hands a working telescope, though without a tube.

PHYSICIST: That was the size of the original Dutch telescope.

HISTORIAN: That's just how my hands are held now; it magnified about three times and it was about a foot long.

PHYSICIST: Of course, we chose the strength of the lenses—that is, the form of their surfaces—to fit that. They were the sort around in 1600, too, standard spectacle lenses of the day.

HISTORIAN: These are optical replicas of the lenses they had in those days. You would find lenses like these, if not quite so good, in the inventory of an average spectacle maker by the beginning of the seventeenth century. According to this line of argument, as soon as the needed lens components became widely available, the instrument was promptly invented.

PHYSICIST: It's quite easy to hit on the right combination using lenses off the shelf. That explains—as you've written—how it is that Galileo could reinvent the telescope in a short time even from rumors and vague word-of-mouth descriptions which were all he had. He made a working telescope of his own in Padua only days or weeks after he had the first news of one that had been shown to the officials in Venice by a Dutch promoter.

HISTORIAN: Certainly, Galileo himself never claimed to have invented the telescope. Once you knew that the working instrument was about, it was really very simple to reinvent it.

PHYSICIST: But the key is that he developed it and used it superbly out of that more or less accidental beginning.

A clock tower in New England examined with unaided eye, then with the concave lens alone, then with both lenses in the Dutch arrangement. This oldest telescopic effect brings the clock dial surprisingly close.

A telescope of this kind, named after Galileo, requires one lens of each sort, convex and concave. The more the curvatures of the two lenses differ, the greater the magnification. By the end of the 1500s, the stock of the spectacle makers regularly included lenses adequate for a three-power telescope.

Galileo in the bell tower. The view of Murano from the tower is seen as the eye finds it, as the Dutch telescope might see it, and as Galileo's tube showed it to the Venetian officials in August of 1609. A senator who saw the demonstration wrote: "I joined many...to see the marvelous and unmatched effects of Galileo's spyglass....One could watch people boarding...the gondolas...where the Canal of the Glass Workers starts."
I too could see exactly that view from the same tower.

HISTORIAN: It is what he did next that makes Galileo so important to science.

Using the lenses we brought along to the top of the Campanile as Galileo did, we examined the same view he had seen. I found it easy to accept that the Dutch telescope with its three-fold magnification was worthwhile for a maritime power. But it is striking to see the ninefold magnification of a telescope like Galileo's improved version, the instrument he brought to the bell tower to show that same view of Murano to the senators of Venice in August 1609.

He did not stop there. He went to twenty-power by November, perhaps to a not very successful thirty-power by January. His telescopes were the first to probe a new cosmos, wherein such instruments have pursued understanding ever since.

What a summer and fall of hard work they must have had, Galileo and his craftsman partner! To make the first set of telescopes, they had to find the best glass, to grind and polish lenses similar to but in detail well beyond the best work of the commercial spectacle makers, and to develop the new instrument to full practicality.

We are lucky enough to have from the collection of the Grand Duke himself two of Galileo's telescopes. They are exhibited today in the wonderful Museum of the History of Sci-

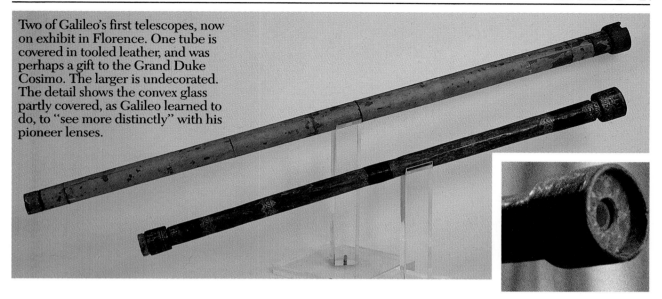

Two of Galileo's first telescopes, now on exhibit in Florence. One tube is covered in tooled leather, and was perhaps a gift to the Grand Duke Cosimo. The larger is undecorated. The detail shows the convex glass partly covered, as Galileo learned to do, to "see more distinctly" with his pioneer lenses.

ence in Florence. One of them is ornamented, and likely to have been a "presentation" model for the duke. The other is more workaday; it may not be the very first one, but its properties suggest that it was an early work. Consider that to make one successful telescope, as Galileo recalled years later, they needed to select among a hundred lenses perhaps five or ten pairs that would work. So they must have made quite a few lenses in that first period.

More than that, Galileo taught himself to use the telescope extremely well, again early in his work. We have his letter to enlarge on that. He wrote:

It now remains for me to tell you what should be done in using the telescope. In brief, the instrument must be held firm and hence it is good, to escape the shaking of the hand that arises from motion in the arteries and from breathing, to fix the tube in some stable place....It is best that the tube be capable of being lengthened a bit, say about three or four inches, because I find that to see distinctly nearby objects the tube should be longer, and shorter for those more distant. It is good that the convex glass, which is the one far from the eye, should be partly covered...since thus are objects seen more distinctly. And so

much can I tell you at present, with all my heart saluting you and wishing you well.

From my house, 7 January 1610
Your affectionate servant
Galileo Galilei

That letter was certainly written to some prominent person, but it was never sent, perhaps because Galileo thought it gave away a little too much of the practical and essential detail before he published the wonderful book that opened our way into the heavens.

His book, *The Starry Messenger*, quickly made him famous. At once he sought out and received the high patronage of the Grand Duke in Florence, his family home, with a higher salary and no teaching duties. Now he was dependent for support on the hazards of royal favor rather than on a stingy but secure contract from the Republic of Venice. Princes do not live forever, and their sons may not favor an elderly scientist, especially one who enters into controversy with Rome; but tenured officials remain in their posts.

WITH HIS FIRST ASTRONOMICAL TELESCOPE Galileo healed the ancient split between the heavens and the earth. What marvels he saw! The crescent moon had shone on the cities of

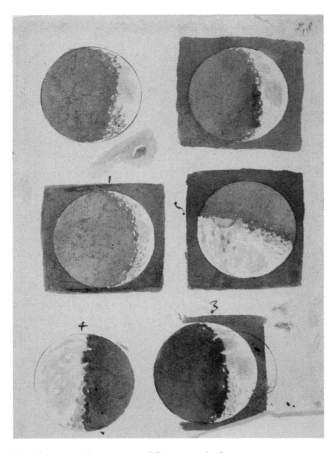

Landscape of a new world, six wash drawings of the moon in ink (and a crater detail) done by Galileo himself, a trained watercolorist. In those days there was no way to reproduce such images; the wood engravings made from them to illustrate his book are very much coarser.

the world for a very long time, and on Florence's domed cathedral thousands of times. But only when Galileo turned the magnifying instrument on the sky were human eyes able to see first of all the substance of the heavens, at least of the moon.

This is what he says of it:

On the fourth or fifth day after a new moon, when the moon was seen with brilliant horns, the boundary which divides the dark part from the light…traces out an uneven, rough, and very wavy line….

There is a similar sight on earth about sunrise, when we behold the valleys not yet flooded with light though the mountains surrounding them are already ablaze with glowing splendor on the side opposite the sun.

He could see the light change on the moon. Of course we can all see it go through its phases, but those changes are imperceptibly slow. The magnified image changed while he watched it, something anyone can see today through a small telescope, as I have excitedly seen for myself.

"And just as the shadows and hollows on earth diminish in size as the sun rises higher, so these spots on the moon lose their blackness, as the illuminated region grows larger and larger." Sunrise in the mountains of the moon! He had placed himself in the geography of the moon. He calculated that the lofty mountains there rose four miles above the lunar plains.

There is another thing which I must not omit, for I beheld it not without a certain wonder; this is that almost in the center of the moon there is a cavity larger than all the rest, and perfectly round in shape. I have…tried to represent it as correctly as possible….As to light and shade, it offers the same appearance as would a region like Bohemia, if that were enclosed on all sides by lofty mountains arranged exactly in a circle.

Galileo had recognized an earthly scene on the celestial moon, and he himself was the first artist of that landscape.

The message that the moon had plains and mountains like the earth was not the only spectacular message from the stars that Galileo read with his telescope. Only a few months before his book appeared, he had noticed in the telescope that next to the bright planet Jupiter there were three faint stars that fell along one straight line. This intrigued him, and he duly followed them for a few nights, to discover to his astonishment that those starlets were in fact companions to Jupiter. They never left the side of that planet; they danced around Jupiter, sometimes nearer, sometimes farther, sometimes hidden by the planet, so that at times one might see four of them, or three or

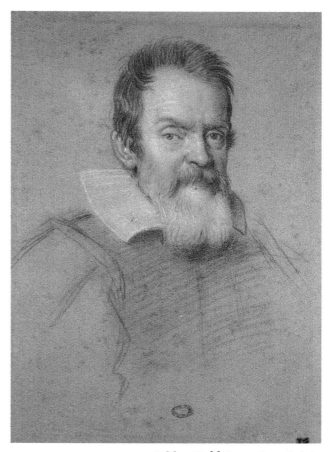

Galileo Galilei, graying at sixty, the portrait done in Rome by Ottavio Leoni.

THE SKIES BEFORE THE TELESCOPE

I T WOULD BE A MISTAKE TO IMAGINE that astronomy began with Galileo, or with the Greeks, or with anyone else who could be named.

Response to the order of the heavens is very old, perhaps as old as humankind. Over the long span of time everything that can be seen today with unaided eye was noticed, even rare events like new stars, comets, eclipses, and unusual alignments of those wandering stars, the planets. Practical sky lore was fully in place, and widely shared. Every country person knew then what few of us know now, that whenever the full moon rises the sun goes down.

The skies were well understood in a different way by the mathematical, who could reckon star and planet positions very well indeed. The accurate star map in metal and the geometrically engraved background plate of this thousand-year-old Arabic sky calculator, called an astrolabe, are witness to a sophistication beyond any simple lore. We can trace the history of astrolabes over 1,500 years from their origins in the eastern Mediterranean through development under the civilization of Islam until they spread as far as England in one direction and India in the other. The points which mark the bright stars on the astrolabe are systematically mapped, each star in its true place on a flattened representation of the northern sky. The position of the star patterns during the night and over the year can be correctly set by the use of the engraved curves on the background, and the place of observation can be taken into account. There was more learning behind the instrument than we can recount here, but the astrolabe suggests the depth of the old geometrical understanding. After all, the annual calendar now in use nearly worldwide, correct to the day over 3,000 years, is entirely a pretelescopic development.

Finally, there was a point of view that

arose both from experience and from the mathematicians, a philosophical picture of the universe. The artist of this superb medieval mosaic in Sicily evoked it visually in all its glowing color and symmetry. The philosophers said that motion here on earth is irregular, unpredictable, shifting, but in the heavens it was serenely circular. Matter on earth is mostly dull, inert stuff, but in the heavens all was luminous. Here on earth we know birth and death, growth and decay. But the celestial world was everlasting.

From creation, then, our universe was split into two, the ideal heavens and the imperfect earth. How could two such different worlds ever be one?

only two. They made a regular procession, an intricate dance he only began to choreograph in the first book. He was so excited by their unprecedented orderly apparitions that he followed the pattern with his telescope wherever he went, even on his trip to Rome. Within two or three years he had recorded the figures of that dance so well that he was able to predict where the moons would be found around Jupiter for some time ahead.

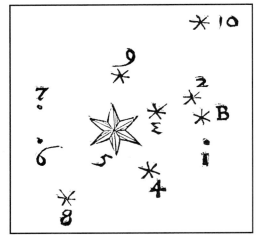

Van Helden and I traveled to a country house upriver a dozen miles from the city of Florence, the old Villa delle Selve, where Galileo had come often as a guest of its wealthy owner, and where he completed his second book on his observations of the sky. We sat on the garden wall and talked of Galileo. What had his early work meant?

HISTORIAN: All of a sudden people lived in an entirely different world. The satellites of Jupiter were a sudden, dramatic, and marvelous discovery for several reasons. First of all, they showed that the earth is by no means the only place that could have a moon. Second, no matter what system of the world you believed in, there had to be more than one center of motion. Finally, they were totally unexpected, a glamorous, invisible novelty.

PHYSICIST: Is that why he named them after the Medici?

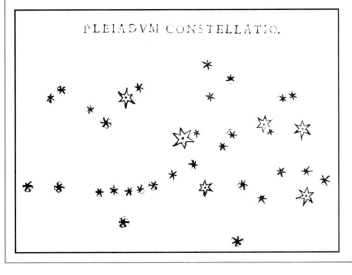

THE FIRST GIFTS OF THE TELESCOPE

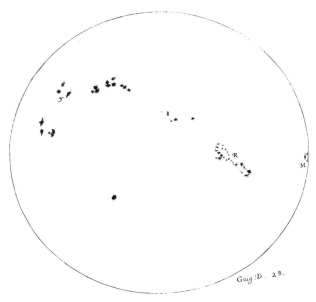

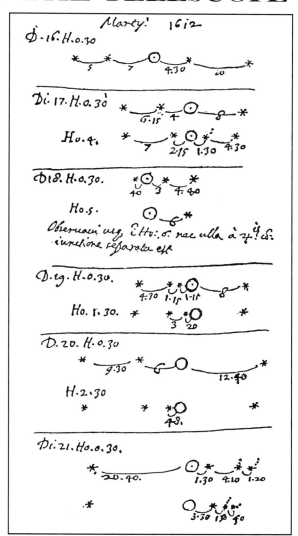

Galileo drew vividly the new heavens he saw through the new telescope: here are the cratered moon and the procession of dark spots that came and went across the glowing disk of the sun.

The star cluster called the Pleiades has six or seven stars for most of us. The smaller star map at the left was made in 1579 by a keen-eyed astronomer who could with unaided eye see eleven stars there. But in 1609 Galileo with his first telescopes counted several dozen new stars. (Modern photographs of the same cluster are to be seen in Chapter 6.)

The diagrams at the right show great Jupiter a circular disc among his little attendant moons, as Galileo choreographed their dance night after night in the spring of 1612.

The two sketches in darker line are taken from Galileo's letters. One shows his views of the planet Venus, whose size in the telescope changed smoothly while its form passed through phases like those of the moon; Venus is always at its largest when it is crescent. Saturn to him seemed an enigmatic triple planet; it was almost fifty years before Christiaan Huygens first made out its famous rings.

HISTORIAN: Patronage, of course, is never very far from the minds of scientists, even today.

PHYSICIST: Any full-time astronomer has to have a sponsor.

But I believe that what counted most of all can be read in one unforgettable sentence in that first book, along with the drawings of the cratered moon. Galileo says that a certain part of the moon resembles the plains of Bohemia. He makes a direct connection between the lunar world and the earth. I think that got him into more trouble and started more speculation, and at the same time took a bigger step for science, than anything else he found. Once you say the heavenly bodies resemble the earth in some publicly visible way, then you are open to pushing the material unity of the world out into the heavens as far as you can go, and perhaps down into the heart of matter.

HISTORIAN: Of course it's not only saying it, it's demonstrating it. The moon's surface is the more accessible discovery. Even with a modest spyglass we can view the marvelous nature of the moon; it hits you smack between the eyes.

PHYSICIST: Were there a lot of people doing that at the time?

HISTORIAN: I think that in 1608 there were thousands of little three- and four-powered spyglasses around already.

PHYSICIST: They were not quite good enough, I suspect, but what about the next step?

HISTORIAN: By, say, 1620, five- or six-powered spyglasses must have existed by the hundreds in Europe. Then the moonscape was readily accessible not only to the scientists but to the imagination of poets; we know that from the literature.

PHYSICIST: The moons of Jupiter are not so hard to see either.

HISTORIAN: No, but you need a bit more power to see them, especially in the seventeenth century with their small lenses, and you certainly need more patience to work on them.

PHYSICIST: The patience is what hurts, if you want not just to look at them once, but to puzzle out their moves. The weather turns bad, you forget, you go away for the weekend....I myself could never really follow any figure of their dance.

HISTORIAN: With the satellites of Jupiter it's the motion that distinguishes them, whereas with the moon it's one compelling look. A sequence or a quick look: that makes a world of difference.

PHYSICIST: It's quite remarkable that he could follow the four satellites well enough to be able to make fair predictions after only two or three years of watching.

HISTORIAN: Galileo was no armchair philosopher, to make broad statements about the universe without real research into the phenomena. Galileo generalized from things that he himself had seen, and he made plausible arguments about the universe based on solid information, just the way we think scientists should do today.

WHEN GALILEO BROUGHT HIS TELESCOPES for the first time to Rome he had a serious purpose. He wanted to show the new heavens in detail to his scientific peers, the Jesuit mathematicians and astronomers at the Collegio Romano. They had read his book, but now they could use the instruments for themselves.

They examined the telescopes, they took them apart, they used them well to look at all the marvels of the sky that Galileo had described. And they announced themselves persuaded by the evidence. In public and private they recorded their satisfaction with the new message from the stars. The princes and the prelates, the dukes and the cardinals, were not long behind. Soon Galileo found himself the center of a glittering company, as the leaders of Roman society wanted to share the new experience.

He could use those times to good purpose. On one occasion, at a great villa on the wooded hill called the Janiculum, on the other side of the Tiber from old Rome, he invited the party to look through his magnifying tube at a familiar church about two miles away across the city. They could recognize it in all detail, and even make out the well-known carved inscription. If the tube reported so faithfully something you knew well, was it not credible when it brought sight of the invisible moons of Jupiter?

At one of those brilliant occasions in Rome, the telescope was given its present rather learned name: *telescopio* in Latin. The English of the day called it the ''optick tube.''

No doubt about it, Galileo's 1611 trip to Rome ended in triumph.

A NEW STRUCTURE TO THE HEAVENS, WITH novelties year by year, does not sound very biblical to us. No more did it appeal to the conservative churchmen of that day, when the Church of Rome was especially embattled in the aftermath of the Protestant Reformation that had crossed half of Europe.

The story is often told. The Holy Office moved to silence Galileo, first by a warning in 1616, and then by forceful demand in 1633. He spent his last eight or ten years under mild house arrest, working until, in the last few years of his life, he became blind. He had to send his final book to be published in a Protestant land, and no longer in Italy.

That outcome is understandable; it is the tragedy of a man, a tragedy of society, the expression of high conflict. It was a time of great division; issues of substance faded away under the more urgent issues of power and authority. But it was not science that suffered in that tragedy. The moons of Jupiter still dance, the planets and the suns still spin, the evidence the telescope brought is plainer still for all to see.

Many historians say that the key issue of Galileo's work was the central position of the sun that ruled the circling planets. For that earliest period they might be right; it was certainly important. But for me that correct conclusion is not what we most owe to Galileo; the proudest legacy of Galileo is the telescope itself. The telescope continues, it lives. The facts it demonstrates, the new experiences we find from it remain. They remain to be explained, they remain to be extended. Where something has been seen wrongly, it remains to be corrected.

There begin the cycles of inquiry and reinvestigation, so characteristic for science based on experiment, on new experience as it comes in. Those experiences do not go away; they increase, they grow. You can ask, Are there four moons of Jupiter, or more? Are predictions of the tables to be improved by one procedure or by another? Are earlier measurements right or wrong?

It is easy to see how this leads to a growth in experience. But the point is that out of new experience come new questions, new explanations, then an extended theory, with new con-

HERE IS ANOTHER EXAMPLE OF THE swift influence of the message brought from the stars by the telescope. Galileo's Florentine friend, Ludovico Cigoli, was an outstanding painter of the time. He was commissioned by the Pope to provide a painting within the dome of the papal chapel in one of the great churches of Rome. Cigoli chose to represent the Virgin. She was by tradition often represented as standing on the crescent moon, shown as a carved or sculptured surface.

This time the beautiful Virgin he painted was familiar enough, but the moon was quite different. It was a telescopic moon, mountainous, cratered, earthy, like the crescent moon in the sky that Galileo and Cigoli had studied together through the telescope. With this unique painting the new instrumental view of the skies first enters the world of the imagination.

The Immaculate Conception, by Murillo.

A NEW MOON

Cigoli's *Assumption of the Virgin.*

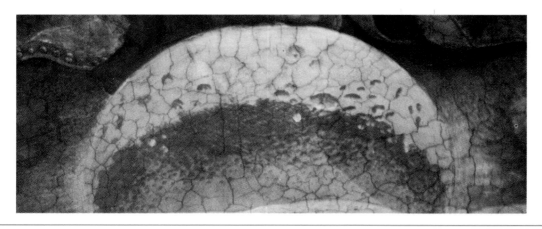

THE WITNESSES

A CENTURY OF TELESCOPIC EYEWIT-ness followed upon Galileo's work. The instruments were improved, they were used more skillfully, and the astronomers deepened their understanding by reflection and experience, the better to interpret the clearer images they saw.

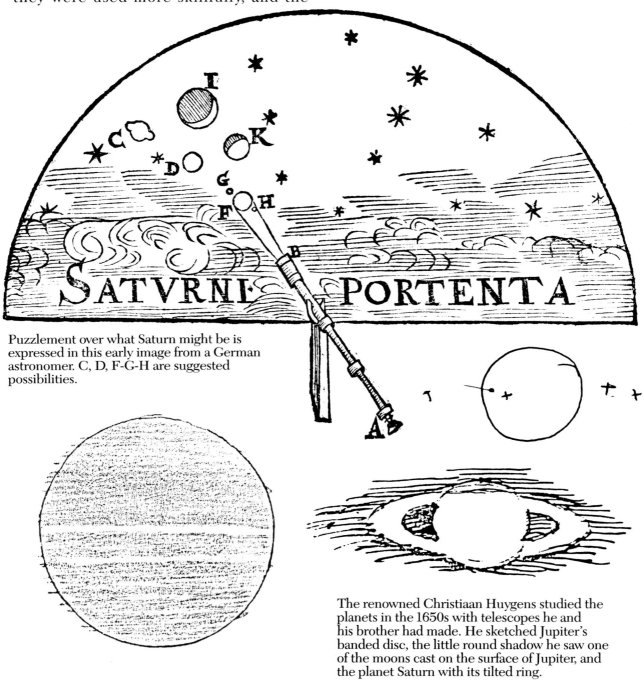

Puzzlement over what Saturn might be is expressed in this early image from a German astronomer. C, D, F-G-H are suggested possibilities.

The renowned Christiaan Huygens studied the planets in the 1650s with telescopes he and his brother had made. He sketched Jupiter's banded disc, the little round shadow he saw one of the moons cast on the surface of Jupiter, and the planet Saturn with its tilted ring.

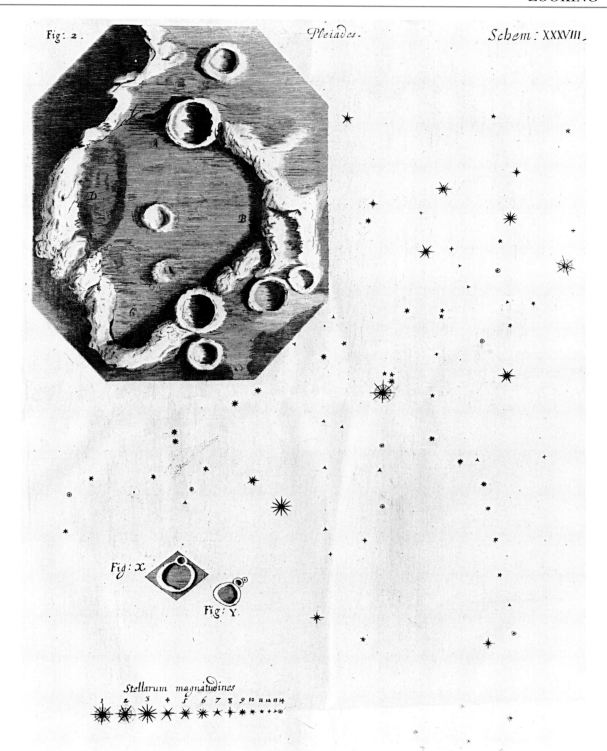

Fig: 2.   *Pleiades.*   *Schem: XXXVIII.*

Fig: X.   *Fig: Y.*

Stellarum magnitudines

Pioneer of the microscope, Robert Hooke published his famous book on microscopy in 1665. He threw in a page of fine drawings he had made at the *telescope*: moon craters, and once again the Pleiades cluster, now six dozen well-mapped stars. Turn Hooke's cluster map on its side to compare with Galileo's version.

cepts. Fifty years after the telescope, twenty years after Galileo's death, there may have been a dozen active observatories in Europe. They had better instruments, able to see more detail, able to find fainter objects, able to measure more carefully, and able to understand much more of what they saw.

That growth has not stopped. I found myself moved by a visit to the garden of Galileo's last residence in Arcetri, among quiet hills at the edge of Florence. The old place is under reconstruction, someday to be used by the astronomers nearby. For just across a little valley full of olive trees, in full sight of Galileo's home, there is a modern observatory, the Astrophysical Observatory of the University of Florence. It holds many instruments, telescopes of a sort that Galileo could not have foreseen but would certainly admire. This is only one of hundreds of places around the world which continue the investigations that began in some open garden or square in Padua in 1609.

Here for me is the confirmation that it was not a tragedy of extinction that Galileo's own work ended in censorship and punishment. The work he began is very far from ended. It has borne fruit in new experience, experience that has become the secure base for a universe whose unity runs wider and whose order lies deeper—though it is still incompletely known—than any that Signor Galileo Galilei,

Academician of the Lynx, could himself have hoped to see.

THE SEVENTEENTH CENTURY SAW THE RISE of modern natural science, beyond the astronomy we center on. It is enough here to recall only a little of what new experience the telescope revealed during the decades after 1610.

Galileo had crudely mapped three dozen stars in the familiar Pleiades. Robert Hooke in London could map six dozen or eight dozen by 1665 or so, and he devised clever mechanical aids to map accurately whatever he could see in the telescope. Christiaan Huygens in The Hague could see with his own better telescope that Saturn was no triple planet at all. He understood that it bore a thin tilted ring, sometimes edge-on, sometimes more open. He first spotted a moon companion to Saturn as well, and decades years later J.-D. Cassini found from Paris four more visible moons of Saturn; the five circling orbits shared regularities with those of the four companions of Jupiter. After several attempts by others, Cassini had greatly improved Galileo's rough pioneer forecasts of the dance of Jupiter's moons. Telescopes now showed the planets as small disks with surface features. From those markings it could be seen that the planets Venus, Mars, Jupiter, and Saturn rotated on their own axes, as did the earth and the sun.

This was a whole new telescopic solar sys-

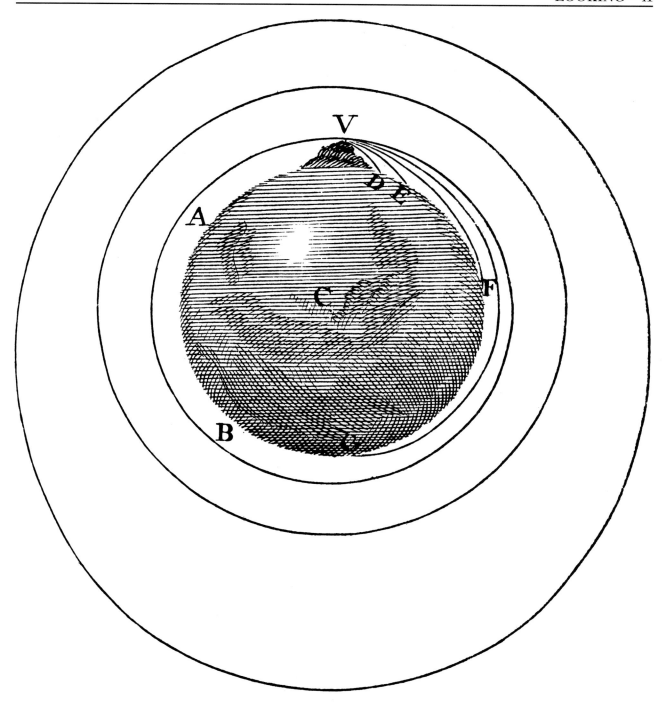

Artificial earth satellites foreseen by Newton
in print early in the eighteenth century. The
substance of earth and the heavens is single,
gravitating universally. Motion enough will turn
a cannonball into a moonlet. (The test was
delayed some 250 years.)

tem, far richer and yet more orderly than the familiar and ancient experiences of the unaided eye, with sun and moon and five planets that appeared only as bright wandering stars. The lawful motions of the principal planets had been patiently charted at last by Tycho Brahe and worked out by Johannes Kepler without data from any telescope. The simpler motions of well-measured new moonlets could be put beside those of the old planets. Along with the orbit of the earth's own moon, and of a few comets, these set rich problems for a great ordering theory.

That order became clear in the mind of Isaac Newton within a hundred years after Galileo's telescope. In his grand unifying books he cited as the first examples of the agreement between his theory and observation not the motions of the old planets, but those of the new.

The century of telescopic discovery had begun when Galileo watched the sunrise on the mountains of the moon. By its end, the genius of Newton had connected even the motions of a falling apple with the measured orbit of the moon. Galileo had recognized a familiar landscape on that distant and silent sphere. That very unity of experience lies at the heart of Newton's profound account of the gravitational order of the solar system. The earth is now to be understood as only one big rotating, gravitating, slightly flattened sphere

orbiting among many others, all with forms and motions at least in part calculable.

Why do we not notice the ubiquitous mutual attraction of gravity that must act between ordinary terrestrial bodies, as it draws the apple down to the earth? Newton showed that it is simply too small to find. "Nay, whole mountains will not be sufficient to produce a sensible effect." It was left to Nevil Maskelyne, the Astronomer Royal to be first to demonstrate fifty years later that a mountain would produce a "sensible effect" with better instruments. His elegant telescopic measurements made clear that an ordinary plumb bob was pulled a tiny distance sidewise out of the vertical by the attraction of a nearby Highlands mountain. Matter, earthly and celestial alike, was all one gravitating stuff.

The universe of science could not be split: visible substance, like calculable motion, is to be found everywhere.

Instruments

PEOPLE HAVE COME EVERY SUMMER FOR fifty years to Stellafane, a green hilltop near Springfield, Vermont, to celebrate the telescope. These are amateur telescope makers, men, women, and young people; all of them have made telescopes for their personal use, or at the least they dream of making one someday. The telescopes they construct are as varied as they can be, but clearly they all belong to that

A day at Stellafane, the amateur astronomers among their home-made instruments.

great family of instrument development that began with Galileo.

For me these people themselves belong to Galileo's extended family. They carry out in backyards and basements with the technology of today the same sort of eager, thoughtful work with hands and mind that Galileo and his craftsman did so long ago in north Italy.

At Stellafane annually there are contests for many categories of amateur work in telescope making and use. One young entrant was standing next to his new telescope with its big wooden structure, the mounting for a heavy glass reflecting mirror ten inches in diameter.

ENTRANT: This is my first telescope, a ten-inch Newtonian. I ground it in....

A CONTEST JUDGE: Your first telescope?

ENTRANT: Yeah.

JUDGE: You made something like this, the first one that you've ever made?

ENTRANT: Yes; I've been waiting for a long time to have a telescope, a ten-inch. Made the primary mirror, finished that in April, had it on the sky last night. It worked beautifully; I was very happy with it.

OTHER TELESCOPES

TELESCOPES GROW AND CHANGE AND yet remain of a family. The two-hundred-inch telescope of Palomar, in California, completed forty years ago, is a prodigious instrument, still among the very biggest, yet Galileo would recognize it from the drawing. The seven-and-a-half-foot mirror that will serve some year soon as the principal telescope in space orbit was ground to shape very much as Don Dilworth worked his glass lens, though at entirely different scale and of different material. The parade of large dishes across the plains of New Mexico is a specialized radio telescope; it appears strange, but it belongs to the family of telescopes nevertheless.

The mirror for the Hubble Space Telescope. The great steel bridge across the top of the picture is the equivalent of Don Dilworth's wooden bar, and the turning-grinding element can be seen at both scales as well.

The Very Large Array of radio telescopes.

LIGHT PATH TO PRIME FOCUS. f 3.3
CASSEGRAIN f 16
COUDE f 30

APPROXIMATE SCALE

R.W. PORTER '38

THE TWO HVNDRED INCH TELESCOPE ~

Plans for the 200-inch Palomar telescope.

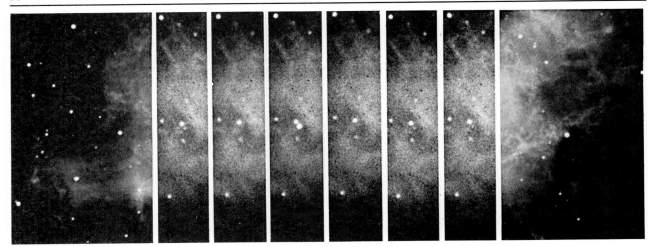

Six narrow strip pictures are flanked for orientation by two wider shots of the outer parts of the Crab Nebula. The strips are stroboscopic views of the Crab, with the twinned central stars in the middle of each strip. Now the extraordinary star is identifiable: the lower right member of the close pair actually flashes on and off twice during the six views. The photos were taken one after another at very rapid intervals; the star repeats the double flash thirty times each second. It has been doing that over centuries, but no one who looked at it had ever before caught its flash.

The Crab Nebula is an enormous luminous cloud visible in modest telescopes, though detail like this shows only in long exposures with big instruments. Many stars are scattered across the field; they are either in front of or behind the cloud, and have nothing to do with it. The pair of very close stars in the center look as though they might be part of the cloud. One of them turns out to be the origin of the cloud: it is extraordinary.

THE EYE IS AN INSTRUMENT, AND ALL THE instruments of vision share some general properties. Here is one cautionary tale of how some powerful modern instruments shared an illusion with the human eye, the consequence of a common limitation.

One of the pair of stars near the center of the Crab Nebula is something of an impostor. It seems an ordinary telescopic star, one of a myriad too faint to see without a good-sized telescope. It has been studied through such telescopes for a century solely because of the luminous cloud, the famous Crab Nebula, that surrounds it. But that star is far from ordinary. Rather, it is all but unique. Its normal appearance is an illusion, an illusion both of the instruments themselves and of a long-held preconception of the observers, amounting to misdirection, an illusion fashioned by no mortal magician but by nature.

The illusion has to do with quickly repeated flashes. Try this: while looking at a picture on your TV screen, flick your hand rapidly across the view, fingers open, looking straight at the screen, hand held not too close to your eyes. The screen will appear to be crossed by darker horizontal bars. The TV picture is actually flashing nearly on and off thirty times a second. Normally the eye cannot catch such fast change; that is the essence of what we see as the flow of motion. But the simple hand-strobing action provides some evidence for the fast brightness changes of the TV screen.

Astronomers never tried anything like that trick through their telescopes: they needed all the light they could get from those faint objects. Whenever they took telescopic photos of the star in the Crab Nebula, they never took snapshots, but time exposures that lasted many seconds or even minutes. Moreover, they all knew well that while some stars certainly could vary over hours or months, no star could flash on and off in a split second. Stars are just too big for such antics. So there was no need to try looking for fast flashes.

But in 1969 they found fast flashes all the same from the central star of the Crab Nebula! In fact it flashes constantly about thirty times per second (only coincidentally at about the same rate as the changes in American television screens). It has been doing so for centuries. Yet no one who examined it ever caught on, because in all that time no one ever looked in a way that could catch so fast a flash.

A special radio telescope had been built in the late 1960s at Cambridge University to look for small variations in the strength of radio sources. It was designed to follow rapid changes, expected because of atmospheric (or ionospheric) effects resembling the twinkling of visible stars. Soon the radio astronomers had found an entirely unexpected clue: somewhere within the Crab Nebula—they couldn't pinpoint where—there was a source of split-second radio pulses.

Then three young astronomers in Arizona realized it was worth looking for fast light flashes that kept the same beat as the radio pulse. Since they knew the exact radio rhythm, their pioneer optical search with fast electronic techniques was quickly successful. The starlight from that one star turns rapidly on and off. But the flashes could not be found at all until we looked with the right instrument.

Our view had all along been partial, an illusion of how we had chosen to look. Like our eyes and ears, scientific instruments, too, can be fooled. Every instrument has its limitations, the other side of its strengths. Different instruments thus may often complement one another. It was the tip from a different kind of looking, one that by good luck had a quicker response, that opened the way to this major discovery. Very small stars can indeed flash rapidly; a fresh way of looking had lifted an old curtain of misdirection.

A few more such remarkable optical flashers are known now, a couple also flash by X-ray, and hundreds of stars have been found to pulse in the radio, all of them tiny collapsed spinning stars no bigger than a mountain, but as massive as our sun.

OTHER SENSES, OTHER INSTRUMENTS

THE TELESCOPE IS UNDERSTANDABLY described as an extraordinary extension of the inborn sense of vision to the distant and the faint. Instrumental extensions of our inborn instruments are fundamental throughout the natural sciences, making a continuity between everyday inference and scientific inference, one that reflects the experience each holds.

Another old example is at once easier and stranger. It is the magnetic compass. A magnetized needle free to turn is of early medieval age in China, where it was invented. Once a few tests are made—Do several compasses concur? Is it some local piece of iron they all indicate?—we can admit that the compass senses something that we ourselves could not detect at all without it.

The next easy lead to follow is that of the rainbow. In its high elegant arch orderly natural processes—the refraction of sunlight through a random myriad of raindrops—have extended the color sense of the eye by freeing a band of hues undiscerned within the white light of the sun. There are several ways to replace the raindrop; consider the oldest experimental device, a prism of clear glass. Newton's prism made a rainbow band of colors; we call it a spectrum. It made objective the subtle personal judgments of

Beyond the red end of the spectrum more invisible light is found. The two images show the eighteenth-century discovery and a modern version: thermometers warm up in the several colors of the sun's spectrum, but most of all beyond the red end, heated by invisible infrared.

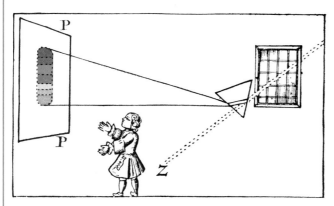

Voltaire brought out an admiring popular account of Newton's work; it had this drawing of the dispersion of sunlight into the colors of the rainbow—the seven "primitive colors"—by a glass prism.

A spectrum indoors made with a movie projection arc. Past the violet end fluorescent letters glow; their surface is receiving light the eye cannot see. Those "Day-Glo" pigments can transform ultraviolet light into visible colors.

color by mapping them geometrically, and it extended them to great refinement as well. But it also disclosed the existence of invisible light, as the compass had shown the quite invisible force of magnetism.

It was natural enough to try what a thermometer would sense in the several hues of the sunlight spectrum, for the heat of sunlight is a common experience. It was unexpected that a thermometer would report that beyond the red end of the spectrum, where the eye saw nothing at all, the thermometer was heated more strongly than in any visible color. We call that invisible color the infrared. This find was complemented at the violet end of the spectrum a year or so later in 1801 by the discovery that a paper coated with silver chloride blackened in the visible hues of the solar spectrum, but black-

See inside a closed wooden box the clear shadows of a chain, a gear, some keys. The invisible X-rays can freely penetrate lighter material. The photo was perhaps the earliest X-ray picture ever made in the United States.

The sky seen by radio telescope, a new starry heavens. Radio intensity is mapped here in shades of gray, using black for zero signal. If we saw this radio band as we see white light, the sky would look rather like this by night or by day. The large white patch is the core of the Milky Way.

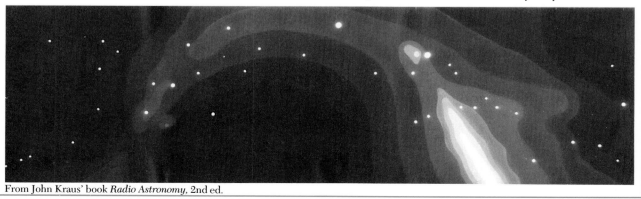

From John Kraus' book *Radio Astronomy*, 2nd ed.

But ours are not the only eyes around; butterflies can see ultraviolet, as can some film. The blue images taken of a butterfly pair by ultraviolet light show the male with strong bright markings, presumably visible to his mate. In our sort of visible light the two sexes look much the same. *Vive la différence!*

ened more readily still beyond the violet end, where the human eye saw nothing: invisible, ultraviolet light.

By the end of the nineteenth century the visible spectrum had been extended much more, both by electromagnetic theory and by direct search with novel instrumentation, to include the radio range, now so important a part of everyday life. A rather more accidental train of events led to the mysterious X-rays, radiation that indifferently penetrated opaque and transparent matter to yield a shadow image on ordinary photographic materials. Radio is also penetrating in a sense, but it did not at once yield visual results. By now, of course, we know how to convert information from almost any channel into a visual image, and a radio map of the sky represents an entire new domain of astronomical experience. The radio telescope is far from Galileo's tube, and yet clearly it is akin to it.

These extensions of the color sense stand for the whole enterprise of the scientific extension of experience through the use of instrumentation. Science is at once parted from commonsense experience and united with it. The sciences are related to ordinary experience as most of us are to the color-blind; there is left only the simple, understandable difference between friends and neighbors.

A moldy tomato holds the principle of a major instrument for the microbiologist.

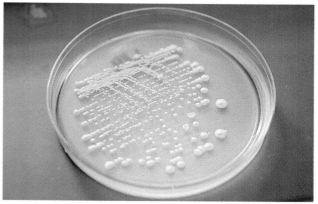

Thriving on the nutrient layer in the flat dish, each spot is a whole clone of living bacteria, all descendants of a first unseen cell put on the plate with others in the sample: now highly visible, easy to count or test again.

ASTRONOMY IS CERTAINLY NOT THE ONLY SCIence, and most scientific instruments are not at all like its brass tubes and glass mirrors. After all, astronomy is distinguished among sciences by its relative passivity. We mostly can only observe the distant world; only rarely can we probe out into it, and almost never can we modify the universe outside.

But out of a homely and somewhat unpleasant encounter with a clearly moldy tomato in the kitchen, there has grown a powerful scientific instrument used worldwide. It is a key instrument for the biologist. Of course, it has no tubes or lenses; its use is not part of astronomy but of biology, specifically of microbiology. That important discipline treats those very small creatures which we know well in sickness and in health, the oldest and the most abundant forms of life.

What can we infer from an isolated spot of furry mold? We can be all but certain that a week or two before there had fallen on the smooth skin of the tomato a spore, a seed as it were, of this particular organism, a fungus. Finding a nutrient situation, just what it needed, it proceeded to grow and mature. At last what had been a totally invisible speck no eye could see, one that only a microscope could find, became not only detectable but conspicuous to the most casual observer.

That procedure in which the scientist takes advantage of a living amplifier, a live organism whose own natural growth turns it from invisibility to easy perception, is the basis of an extremely powerful microbiological technique.

Of course, the flat round covered dish, called the Petri dish after its inventor, is not so indispensable for the technique as is the lens for the telescope. But it is useful, and can be seen by the hundreds and even thousands in any modern microbiology laboratory.

Petri dishes are in heavy use here.

The scheme is almost as simple as the tomato. Prepare a nutrient gelatin, often clear and transparent, in the Petri dish. Usually you would sterilize the whole dish in live steam, to make sure that no viable contaminant is present. By choosing the nutrient recipe, as well as the temperature, added drugs, and so on, you can favor the organism you wish and inhibit the others; tomato molds do not grow well on an orange skin. Microbiology is always interventionist.

Now spread your sample on the surface, cover, store as you wish, and wait for growth. If the sample you added had even a few (but not too many) of the living organisms expected, you will see that each one has been succeeded by a distinct colony of its descendants, each colony a clone of its ancestral cell. A search by microscope through a teaspoonful of sample would have taken a very long time, and would not show viability. At once this technique has made pure clones, which can themselves be propagated anew to find the properties of the creatures. A Petri dish of colonies is like so many plots on an experimental farm.

It is extraordinary how ingeniously and how far this broad method has been developed and extended, first in microbiology and now within molecular biology as well. I have no hesitation in saying that this idea of a pure culture as living amplifier is a major instrument for microbiology, if only symbolized by the little Petri dish itself.

Hurrah for the instruments then, all of them!

Some are the instruments we inherit from birth and know well how to use—the eyes, the ears, the active probing touch, in general the human senses. Some of them are the changing and remarkable instruments that come out of science and technology. I hope I have shown that those two classes of instruments are deeply related; the lens of glass is simpler than, and yet not so different from, the living lens of the eye.

Expositions of science rarely put scientific instruments at the center. The writers focus rather on the concepts and the findings, the high results of science. Those results are the building blocks of the house of science, made of experience and explanatory theory. But the instruments are the builders' tools and means; they quarry the blocks in the first place, they level and fit them.

Constructing science is even more intricate than that. For a theory we have had for some time is almost bound to fit. It has been tested many times against the same old experiences. Its edges are well worn. Given such a past we seek something new; perhaps some strange new image is what we need to challenge the validity of any theory. The instruments then become more than just the means, they become the new searching tests, the guarantors of our understanding.

Of course, new theories, new findings in

turn bring new instruments. These in time probe the very foundations on which they themselves rest. Once the radio was only a new result, a marvelous prediction of the electrical theorists; once X-rays were another dazzling result of a chance discovery. By now the two of them are invaluable tools of science, providing instruments for the astronomer that show quite new faces of the universe.

It is a long exciting round: old questions, that after a while receive hard-won answers, then themselves evoke new and sharper questions. It is the new instruments and the new experiences they bring that sound that rhythm loudly for every living science.

THE MULTIPLE IMAGE

THE EYE AND THE MIND CONTRIVE A smooth flow by sewing together the repeated samples they receive from the moving world. But both the scientist and the artist have found a way to suggest motion and change without that internal process we so little understand, in a way by taking flow apart instead of putting it together.

Here are three instances. The first is the work of a technical pioneer, photographer Eadweard Muybridge. He spent a lifetime in the study of motion by the use of many still

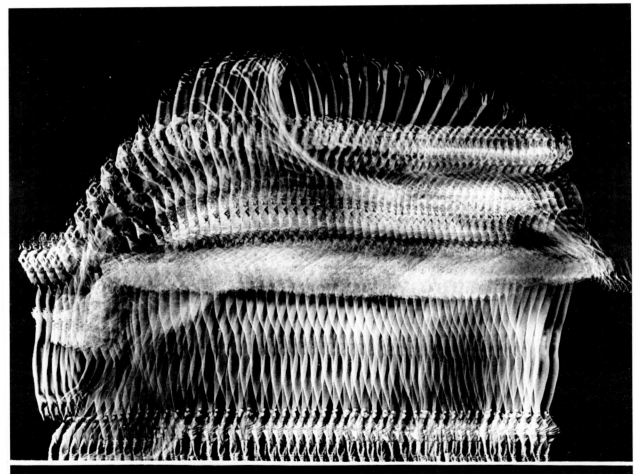

Gjon Mili, 1947.

Eadweard Muybridge, 1887.

cameras set in a row to watch, their shutters triggered one after another by trip wires as the subjects sped by. His comprehensive work is as beautiful as it is instructive; it unites science and art, and certainly it was a forerunner of the cinema.

The painter Marcel Duchamp used the same sequence of images as the basis for painting motion; it is tempting to think that he had seen Muybridge photos, cinema film, and animated movies, but his was so fertile a mind that the influence might not have been necessary.

The photography of Gjon Mili came later, once the flashing strobe light made it quick to do what the other two had carried out laboriously.

End here in one diverting frontier between science and art, where both have extended the ways we can look at this inexhaustible world.

Marcel Duchamp, 1912.

The Philadelphia Museum of Art: Louise and Walter Arensberg Collection

2

CHANGE

EVERYTHING FLOWS, SAID THE PHILOSOPHER; no condition is permanent. Ours is no static world of forms and structures alone, but a world of events, of change; there is both being and becoming, to echo the fine old phrases once more.

Let a high cliff of ice collapse or a feather fall, let a sculptured sheet of steel rust over the decades to a red-brown mass, let a flower grow or a hard kernel of popcorn burst into white softness, let a whole forest burn or only one single matchstick smolder: within every change there has been found something that is unchanging. The transformations can be large or small, fast or slow, extraordinary or inevitable; we believe with reason that they all satisfy the rules found to hold within every diverse process that has been checked. One of the most unexpected and powerful results of all science is that we have teased out of every event a quantity or two that remain faithfully unchanging and constant throughout the widest transformations.

That grand law of constancy is the topic we now pursue. In the jargon of the day, it is the universal bottom line, the one verifiable account that in our experience has never failed to balance.

What the Eighteenth Century Found

EVERYONE VIEWS A SPARK AS SOMETHING ephemeral, vanishing. Sparks are a symbol of the insubstantial. They suddenly rise, glow for a little time, to die almost at once. That is a naive view of the world of change, a viewpoint that depends upon the fact that we are visual creatures, who fix too quickly upon just what we see.

It is easy to catch some of the sparks that flood out from an ordinary Fourth of July sparkler. You spread a kind of net, simply a sheet of ordinary white paper. The sparks fly out, only to disappear. A few bounce on the paper just before they die.

Has the paper caught anything lasting? Simply tap the paper gently, after partly rolling it into a large *U*, so that anything that fell might be collected. A dark streak of material certainly appears at the bottom of the half-rolled sheet. It may not look very beautiful, but it was not there before the sparks flew from the sparkler. Now one asks a lens to aid the eye, an ordinary magnifying lens.

Under the lens there is something quite astonishing. Here is a host of small, glittering, smoothly rounded beads. That is what is left in

From the popcorn kernel that expands to a white puff within a few strobed flashes to the slow conversion of shaped steel into shapeless red rust, the world is full of changes, great and small.

A sparkler and its sparks, first seen as they momentarily dazzle the eye, and then after they have left easy view to become beads of lasting iron.

the streak. The sparks have indeed not gone; they have not disappeared, but left a residue of some beauty.

Of what are they made? We know an easy way to ask. Slide a small horseshoe magnet along the streak. The little beads love it, and nearly all of them leap to it. The fireworks maker would not be surprised; the artisans who composed the mixture of the sparkler put iron or steel filings, scrap material, into their mix. The rapidly burning fuel quickly heated the filings; as molten droplets they flew through the air, formed as round as can be. The bright sparks we saw are such droplets, which scattered, cooled, and disappeared from sight, but not from the world.

The sparks were physical objects, tiny beads of iron, first molten and glowing, shiny metal once cold. It is plainly very dangerous to over-indulge the eye, to assume that because we no longer see something conspicuous, it has ceased to exist. Here material is not lost, but conserved.

TRY ANOTHER EVERYDAY EXAMPLE. A CANDLE flame is bright; around it we expect really nothing but light and heat, and the tremulous rising current of air. Nothing more can be seen. But for a long time curious children have managed to reveal something around the flame that is by no means visible. It takes no more apparatus than a cold spoon. Place the spoon against the flame; quickly a kind of image of the flame appears drawn out in black soot on the metal surface.

That is certainly matter which has condensed there. The test can be tried in a modified way. This time hold a cold knife blade a little farther above the flame, and move it through the right place just above the visible tip of the flame. A breath of steam will condense on the knife blade, to vanish almost at once as the metal heats up. Water, too, is there within and from the flame. The flame is a source of substances that do not end when and where the light ends. The invisible must be taken into account; it may be as significant as what we see.

After all, why should the processes of the world pay special attention to the nature of our vision?

Common speech and common sense are firm on the usage: the clean glass on the shelf is empty. We know that it is full of unseen air, but the commonsense experience of filling it with water by pouring from above pays no heed to the air already there, which finds its way out easily. But try to fill the glass in a different way, say upside down into the bathwater, and you recognize at once that air is present and not at all to be ignored. The diver who makes her way underwater to the emergency air provided in an inverted plastic bell regards that vessel as full of the most precious stuff, even though its content of air is not to be seen any more easily than the air in the glass on the kitchen shelf.

Open the hood of any car. The running engine draws in air, sometimes through the large air filter, often through a special tube only inches across that extends to the very front of the space underneath the hood. Dangle a sheaf of ribbons of filmy plastic, or hold a source of smoke just ahead of the end of the intake tube, and gun the engine. As the engine roars, the ribbons or the smoke are drawn strongly inward along with the air flowing into the tube. The engine is breathing hard, and that tube is its nostril.

Fish, insects, birds, mammals, and humans all breathe, just as every fireplace requires air, just as all auto engines do. Without air, they all perish. It is natural for us to overlook the auto's urgent need for air. Whoever drives a car must attend to and pay for the input of gasoline, but the equally essential input of air, really a co-fuel along with the gas, is free, and we safely ignore it. For the air flows to lung and engine from the ambient atmosphere quite invisibly. The designers know that need, and provide for it, often very simply. Some engines, designed to produce extra power, demand more air than is made available by casual flow, and instead pump it in avidly with a turbocharger.

Living forms continuously breathe to gain air, a noticeable action. Ancient physicians did not know the purpose of our breath: did it cool or stir the blood?

It may help us grasp the extent of the role of air to note that for every gallon of gas an engine uses it must also consume all the oxygen in more than a hundred pounds of air. That is the volume of air that fills an ordinary room, all invisible. Each gallon of gas burned takes up the same amount of oxygen that a dozen people require in a whole day.

The diver knows and counts on the fact that a vessel is not at all empty when it is full of air; the description of a glass on the shelf as full of air is no longer academic once you expect that the glass might be encountered in a variety of conditions, not just on the counter near the sink. Add to that the recognition that our inherent senses—like any means of perception—are partial, and you find the need to examine any situation not only by a glance of the eye, but close up, magnified, by chemical analysis…on and on. A deliberate extension of common experience is often the essence of a scientific experiment, built either on a new instrumental perception or on a new context for what was otherwise familiar. Our daily understanding of our actions is not that far from a scientific account. The difference is mainly that common sense assumes a commonplace environment and ordinary circumstances. But the scientific account is designed instead to work under a large variety of circumstances; it is tested against far wider experience.

THE NEXT STEP BEYOND THE COMMONSENSE emphasis on the visible is to introduce an old

Burning matches lose substance and hence weight. The balance tilts.

and familiar instrument for the comparison of samples of almost any substance: the balance. It is in steady use (perhaps in modern modified form) at the supermarket, certainly at the places where they pack and label the boxes and tins you buy. The archeologists have excavated a workable equal-arm balance that was in use in predynastic Egypt, six thousand years back.

The first sign of the constancy during change that we seek was disclosed by the balance. Gradually the use of that instrument spread to craft after craft as part of an artisan's specialized knowledge. The general result was made formal and recognized as universal by laboratory chemists about the time of the French Revolution.

The flavor of the result can be caught from a few very easy applications of a good balance. Suppose you balance the weight of half a dozen kitchen matches placed on one balance pan, and set them alight. They burn brightly; once the little fire is over the pan that holds the charred matchsticks will swing upward, much lightened. Everyone knows that wood burns away to a much reduced bulk of charcoal and ashes, and the balance concurs. The invisible products of combustion, and smoke, sparks, and soot with them, have taken away weight as well as bulk.

Not every change loses weight. For a contrast you have only to place a spoonful of a white powder commonly used as a drying

Bright red, the compound mercuric oxide decomposes upon gentle heating into its two elemental constituents, the clustered droplets of condensed fluid mercury and the invisible oxygen. Sealed in glass, the weight does not change, though the appearance changes sharply.

The pioneer chemists of the eighteenth century learned how to account for the invisible air as an important chemical substance. This drawing from 1775 shows one of the gas-handling schemes of Dr. Joseph Priestley, the discoverer of oxygen—and of soda water.

agent on a sensitive balance. Within a few minutes the weight of the powder will have increased, and it can be seen as well that the once-dusty stuff is becoming moist. Out of the invisible vapor in the air liquid water has come to join the dry powder.

Finally, consider a more active experiment. Again, at the beginning place matching objects on the two pans and balance carefully. What is put on each of the pans is a small bottle of water with its cap off, that cap itself, and two tablets of the famous Alka-Seltzer (pressed discs of sodium bicarbonate and citric acid): equality.

Now compare two differing procedures. First drop the tablets on one pan into their bottle of water, and do no more. The famous fizz appears, the bubbles rise and invisibly depart, and that pan rises, definitely lightened: loss.

For comparison the same two balanced pans can be set up exactly as before. Again put the tablets on one pan into their bottle, but now quickly screw tight that bottle cap, sealing the bottle, and at once replace the capped bottle,

pills inside, on its pan. The bubbles again rise furiously until the action stops. But now the pans of the balance do not shift; the balance is maintained: equality again. A simple cap is enough to trap the invisible, and the weight is unchanged, although the solid tablets have vanished into fizz and solution. (I think you have a clear idea of what to expect when that cap is opened: pop.)

An old piece of farm machinery rusts for years in the abandoned field. It is gaining weight steadily as the metal takes up the invisible corrosive reagent, oxygen from the air. That might be checked over the years without much trouble—provided that the rust itself has not chipped and fallen off into the soil as wind and weather beat upon the relic. The speed of the change is of no importance to the balance.

A Box of Changes

HOW GENERAL ARE SUCH RESULTS? CAN THE balance remain during all sorts of marked change in matter and form? We prepared a playful and entertaining version, real enough, of an experimental test that has been done many times, much more slowly and much more soberly. The idea is simple though deep. Seal a big container tightly enough so that neither invisible gases nor iron sparks nor any other material can leak in or out. Arrange for a wide variety of changes entirely within the self-contained box.

I began by thinking of it as a boxful of verbs: to move, blow, spin, fall, explode, swell, ignite, glow, burn, buzz…. The whole box is supported on a suitably sensitive balance. The aim is to see how the total weight of the box changes during all the variety of internal changes mustered before our eyes.

Tom Tompkins, the amiable and ingenious engineer who designed the box for us, had solved a set of agreeable problems. The box had to be physically well sealed, quite self-contained, so that no material went in or out, and no tubes or wires either. That meant that the electrical power sources had to be internal batteries. A motor turned a rotating set of contacts that switched on one device after another. Almost all of them had to be set off electrically, including candle, sparklers, and firecrackers. The technical art of electrical candle lighting is not well developed! But it all worked gloriously.

The entire chain of events was initiated by a little magnet outside that rested on the lid of the box. Once I slid it away from the spot where a steel ball bearing had been held poised in place within the box by the pull of the magnet from without, I had started a chain of events in the very style of Rube Goldberg's marvelous old cartoon contrivances.

Moving the magnet had triggered a string of funny little devices isolated within the big sealed box of transparent plastic. One after another the enclosed changes proceeded: a candle glowed, battery-run toys moved and chattered, a layer of popcorn kernels flew up to burst noisily, sparklers burned, a magician's "spring flowers" bloomed, their springs expanding as the catch was released, a radio began to play, the wind of a whirring fan spun a

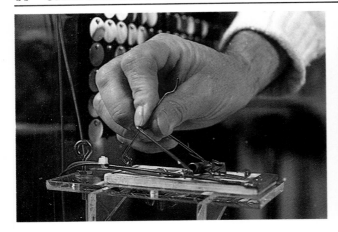

The box of changes encloses a variety of events, unwinding, battery discharge, motion, flame, light, sound, explosion, falling, spinning…. The box is sealed well, as the smoke shows by its cubical shape. The old farm balance with its load of bricks does not change by so much as the weight of one twenty-five-cent piece.

cheerful pinwheel—plenty of change—until, as a climax, a whole bundle of firecrackers went off in a ripple of bangs, and still the candle glowed through the dense and wholly contained smoke.

Rube Goldberg's cartoon legend would have read something like this: operator moves aside Magnet A, so that Ball B falls upon the trigger of Mousetrap C, which snaps closed, to switch on all the action in turn, from D to Q!

When all was over the smoke filled a neat cube; that offered visual evidence that this box was pretty tightly sealed, and nothing like smoke or sparks had leaked. First of all we had balanced the box against bricks and coins on an antique farm scale, once used to weigh sacks of grain. We also checked the response of the balance; a change in box weight by so much as the weight of one twenty-five-cent piece would show up. The box retained its balance very well, the same after as it was before the dizzy set of changes. The weight did not change, as judged to better than one part in a thousand.

All the changes we saw, from the steam-expanded popcorn, to the discharging battery, the burning candle wax, the blast of moving air from the fan that drives the pinwheel, the motions and sound of the little toys, the expansion of the springs, the radio music we heard, the glow of the sparklers, and the final smoky explosion of the firecrackers—for all those mechanical, chemical, electrical, visual changes in position and state, nothing changed the balance.

Weight—that which we measure on the balance—remains constant during the widest variety of changes. That is what was shown in this rowdy version of an elegant experiment that has been repeated with great care ever since the eighteenth century. The result remains a pillar today throughout science and technology, in lab, shop, and factory.

As we removed the dozens of screws that held the rubber-gasketed lid tightly to its box, the smoke jetted out neatly through each newly opened screwhole. There was increased pressure inside the box, for the air had been heated by all the action within. That pressure sets a practical limit to the amount of change we can allow inside a sealed box, especially when we require a box open to easy view from outside. Tom Tompkins was well aware of such a limit. It had been one of his chief concerns: how to make spectacular changes and yet not blow open the box. He was pleased to see that the overpressure had been contained, and yet was noticeable upon opening; he had not grossly overdesigned his tight box and its lively contents.

WE HAD WEIGHED OUR BOX OF CHANGES TO the accuracy of the weight of a twenty-five-cent coin; that was the best our big old scales could do. Weighing can be done with remarkable

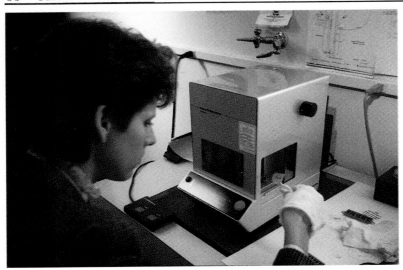

Loading an ultramicrobalance to weigh a small dot.

sensitivity, determining very small weights; I tried it to gain some feeling for its limits.

Clean white gloves and tweezers, used in a very clean lab, are *de rigueur*; otherwise the fingerprint marks and skin dustings we can't help leaving behind will make our weights variable, noisy. I had the generous guidance of an expert technician who supervises a small modern balance room. Her balance is set up to monitor particulate air pollution at many sites by precise weighing of the small smudges deposited by air currents drawn day after day through strategically placed filters.

The windowless room had lab benches, but the ultramicrobalance we used did not rest on a simple lab bench. It resides on a special granite tabletop, mounted on shock absorbers that sit squarely on the concrete floor slab. Vibrations add noise. So does static electricity: very light objects feel little gravity, and it is important to eliminate any stray forces on the samples. Devices that neutralize static electricity are mounted here and there around the balance case. The stainless-steel balance case, a foot or so on each edge, has big clear windows so that we can see what is going on inside.

The reading is electronic: the shifting digits glow. This balance has a single pan about an inch across. An automatic system swings the pan out into an antechamber where you can reach inside to place the sample, once you lift the glass door. But before the sample can be weighed that door must be closed, and the motor turns the pan you have loaded back into place, through a lock that opens to admit the pan and closes again, to shield the weighing chamber from the dust and breezes of the outside world. All these actions require pressing the right control.

An electronic scheme senses very small deflections of the main beam of this balance, a beam hidden from view within. Its makers have made the device capable of recording weight differences down to a tenth of a microgram. (The prefix *micro-* means one-millionth part.) That is one part in a few million of the maximum load the balance will accept, which is about the weight of a dime, two grams and a half, just under a tenth of an ounce. I had never used so delicate a balance; in some research laboratories, balances that use gravity may have gone tenfold better than this one for a comparable load, but hardly better than that.

I wrote my first name in pencil on a small slip of roughened white plastic. (The weight of a piece of paper is so much affected by the changing humidity of the air that we were advised against its use.) The balance accepted the slip and gave its weight, in the range of a tenth of a gram. But we were after the weight of just the dot over the letter *i*. So we reset the reading to zero. Now when the balance reweighed the dotted sample, we read out directly only the difference, the added weight of the dot (plus or minus any dirt we might inadvertently add or lose). The signed slip came out; I held it with gloved hands and penciled a rather emphatic dot over the *i*. The balance received the slip, and reported that my awkward dot weighed 162 micrograms.

That was a clumsy, heavy dot as dots go; trying a few more, we found that a minimal, inconspicuous sort of pencil dot weighs five or ten micrograms. My full name weighs in at the thousand-microgram range.

This experience gave us some appreciation of what a microgram might mean. It is not much stuff (although we will find it is a myriad of atoms). It is not a great trick to weigh tinier samples down to a thousandth of a microgram, I imagine. But handling them cleanly then becomes the major challenge. The balance delicate enough to weigh a thousandth of a microgram can hardly accept loads as gross as one gram.

We can weigh down that far perhaps, and we can detect changes as small as one part among five or ten million. That is as far as we can go in extending to smaller weights the inborn sense of hefting two envelopes, one in the palm of each hand, to judge which is heavier. We can all do that to a few parts in a hundred.

About the turn of the century our box of changes was anticipated by an extremely demanding test. The changes studied were chemical reactions between solutions; their sealed containers were glassware specially made for the task. (It wasn't easy: the heat of the reactions caused minute amounts of water to evaporate from the glass surface until the trouble was noticed, and the glassware wax-coated.) The whole setup weighed some fifty grams. The best measurements with the non-electronic but delicate two-pan balances of the day were good to a few micrograms, say a part in ten million. No change in weight was reported before and after marked chemical change, to that degree of refinement.

Weight is constant during every change as far as we can tell.

We like to think of our complex living world as open to the winds that blow. In a way it is not; on a larger scale we all dwell in a closed world.

Some clever biologists have lately made little versions of the closed world of all life. They gave us a few of their little worlds of complex living things, called ecospheres. At first sight they looked rather like goldfish bowls. But they are not at all the usual fish tanks; they are breathtakingly different. These are tightly sealed globes, separated from any material input or output by the permanently fused glass which encloses them completely. But they nourish the cycle of life.

The tiny red shrimplike creatures that swim in the salt water amidst the green algae cannot

Two sealed spheres, closed ecosystems. The big blue earth-sphere is all but sealed by gravity within its rarefied space environment; the small sphere is sealed by fusing the glass. The animal inhabitants of the little sphere can be made out as red streaks almost at the left edge.

be fed from without. No air at all enters or leaves. Water cannot be added or removed; the bowl can never be cleaned. The fate of this life, whatever it will be, is sealed within its walls of glass. The material balance that the organisms strike within, the cycling air and water and the creatures that live and die within, will be what they will be as the years pass. We know that the account is strict. Nothing material will come in, nothing will go out.

Certainly such a sphere depends upon an input of sunlight, upon the rhythm of day and night, upon the ambient warmth given by our own atmosphere. In that sense the spheres are tiny distinct planets, each one a toy earth. What one group of organisms discards as waste can and must help maintain another group. It is to be realized that a great many microorganisms are invisibly present that contribute mightily to the chemical interchanges within, a remark true as well for the big blue earth.

Is our own life so different? On Planet Earth, there is no glass wall sealing our atmosphere high above us; instead gravity holds it in place. Our earth too has a material inventory that is almost entirely fixed. We derive essential sunlight from the same distant place as does the little sphere. That provides the warmth we need and the rhythms. A little material comes in to earth, like meteors and meteoric dust; a little goes out, the slow leak of light atmospheric gases and an occasional space probe.

Matter comes to the earth's surface from the deep interior and is buried again. Even that is only an internal rearrangement. For all essentials, over times as long as the history of human life at least, our earth is a closed system like the spheres.

Like the ecospheres, earth, too, has a fixed material balance and a constant weight. We have not measured it precisely, but we firmly believe its weight remains all but constant. Everything that happens to fix our fate, certainly over millions of years, is but continual rearrangement, the constant ebb and flow of all the materials that pass ceaselessly from creature to creature, from place to place, from form to form. That internal balance is guaranteed by the constancy of weight through change.

AN EXPLANATION FOR THIS REMARKABLE constancy of weight during change had been waiting in the wings for a long time.

Suppose it were true that the gunpowder and the air and the chemicals of the storage batteries and the metal of the springs inside our box of changes were all made of little modular particles. Let us call them "atoms." And then suppose that what happens in every change, a change from wood to ash, from gunpowder to the products of explosion, or in the discharge of a battery, is nothing but a rearrangement of those atoms, a complicated, intricate rearrangement down deep within substances, the parts too small to see. Those parts are all there, the same parts, just as many as before. They have only rearranged themselves, in an unending pageant of dances and assemblies fast and slow, so that now they hang on to each other in different intricate patterns.

You could imagine that. In fact, it is called the atomic theory of matter, and as a speculation it is very old. It automatically explains the constancy of weight, because from experience we know that just rearranging the position of the weights as they stand on one pan of a balance will never disturb the balance. That is far from proving that atoms exist. No, no; we could not claim that as proof. But it surely brings cheer into the hearts of any who seek an atomic theory before it has yet been demonstrated in detail.

Theory or no, the decisive point is that there is a material balance. However substances and forms of matter interact and transform, there must remain guaranteed full weight. Just because you ignore waste products and wish to throw them away, your intention does not arrange for any matter at all to disappear. Whether down the drain to the river and the sea, or through the exhaust pipe or up the smokestack to the open air, it all goes somewhere. It retains the same weight wherever it goes, no matter how widely it may spread.

We have watched a simple test in a closed system. The earth, too, is in effect a closed sys-

tem. All its material cycles around the earth, to be used, reused, stored, modified, but never to be lessened or lost, just as within the little glass spheres a few visible plants and animals and their progeny live their lives among unseen microbial partners with a strictly fixed inventory of material.

That constancy of weight throughout all change, and the controlling applications of the result in almost every technological and scientific discipline, gave the first sign of such a powerful discovery in science.

Yet it is not the only example.

What the Nineteenth Century Found

FOR ITS SEVENTY-THIRD YEAR, THE CYCLISTS of the Tour de France raced about 2,000 miles in a grand road circuit of the whole country, starting near Paris and ending in that city twenty-three days later. It is the most famous endurance race in the world, a sporting event that transfixes French attention, as television brings it every evening to the nation.

The 125 racers of the Tour spend six hours each day racing, with a traditional single day off in the twenty-three. There is no stopping on the road; even injuries are treated at racing speeds. Lunch is passed on the run, as well as the water they so much need. For days they speed across the flat countryside under the July sun; for days they wind through the Alps and the Pyrenees. The mountain road may

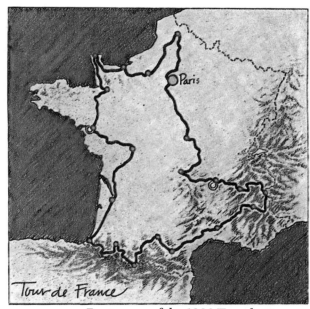

Route map of the 1986 Tour de France.

climb two miles in a single day, then down from the Alpine pass they spin at seventy breakneck miles an hour. They average about twenty miles an hour over the whole course.

Sometimes a rider is alone on the road. Sometimes he is bunched into the pack, where twenty-one teams contrive their subtle tactics. The overall winner in speed gains a princely prize; most of the professionals expect to finish only an hour or two behind the winner. This is the grandest feat of human endurance in the world.

We waited at roadside for the 1986 Tour in rolling country in the south of France, about halfway. "There they come!" I shouted in excitement. Fourteen days, every day racing on the roads, and still they were making twenty miles an hour. Tens of millions of people are caught up in this wonderful race. They are eager to know who wins, and maybe who is second and third. But I am not much interested in the winners. I know nothing about winning such a race. Seventy years of experience, intelligence, passion have gone into the endeavor: tiny differences decide it all, and an outsider can only admire the intricacies.

For me, these racers are all winners: the first hundred riders who finish the race all chalk up above 95 percent of the speed of the top performer. That is what is extraordinary, for this is a remarkable physical performance, possibly unmatched anywhere else. I can suggest its phenomenal quality for you in one simple way: Each day each racer in this Tour de France puts out two or three times the amount of physical work done by a runner who completes a fast marathon run of twenty-six miles! Then the cyclist does it again the next day, and the next day, and the next, and again....

We can throw some light upon this feat by the curiously impersonal but very deep methods of science. That is why I came to examine the Tour de France. We can find some order in the Tour de France if we invoke what I think is the deepest insight of nineteenth-century science, rather as the recognition that weight is unchanged through all sorts of processes had been a magnificent finding of the eighteenth century.

Everybody knows this second one, at least by name. It was the identification of the idea of *energy*. Energy is not a substance, visible or invisible. It is an intangible, abstract measure that applies to many forms of change, to motion, light, sound, magnetism, chemical reactions like the burning of fuel or the digestion of foodstuffs—all of those changes, and many more, can be described as the transformation of energy from one form to another.

Energy flows freely from form to form in the processes we watch. Transform it as you will: once you learn to take account of its flows, you find that it never disappears and never appears. It may enter or leave, but it is always accountable in full. It never vanishes, it never comes anew out of nowhere. That conclusion was the work of the nineteenth century, jointly established by chemists, physicists, physicians, biologists, meteorologists, geologists, engineers.

Given that secure finding, let us try to look at the Tour de France from the point of view that sees a racer as a system which takes in energy and must therefore either retain it or give it out, or some of each. Nothing can be lost.

In the first example we gave of the search for constancy with our balances and our smoke-filled cube, we undertook a simple test of unchanging weight. Here I do not intend to check the validity of the second great result: energy, like weight, remains constant in every change. Rather I will *assume* that it holds, and instead show how important such a principle of conserved quantity can be by directly *applying* it to the bicycle race.

We will find out how much understanding it gives, how tightly it sets limits and enforces compromises, even when as here we know only in small part the detailed processes within the human body and the chances of the road. It would be unwise to try to verify a result so amply demonstrated within a complex set of changes that wind along with the speeding cyclists across a whole applauding country. But use it we certainly can.

WE BEGIN WITH SOME VERBATIM TESTIMONY from the Tour racers themselves on a personal matter. "What do you eat for breakfast?" they were asked. We cite informal fragments of the offhand answers, given in many a tongue by these professional athletes:

"I have a Hostess Twinkie before I get up."

"Start off in the morning with a, like a continental breakfast."

"Café, petit pain, confiture…"

"Oats and rye and all those sort of natural, natural things…"

"And then we have some steak, rice, muesli…"

"And then we go for some spaghetti, rice, chicken…"

"Les oeufs, spaghetti, la salade de fruit…"

"Well, basically I just have muesli, where most of the riders have spaghetti…"

"Maybe an omelet instead of some chicken, but you know, some protein…"

"An omelet, say, but I just stick to muesli and bits of bread, you know?"

"Couple croissants, good fat content…"

"And we have muesli, oatmeal, you know…"

"Have a bit of coffee, get the heart going…"

Later in the morning, a little before the race:

"We get out here, have a peach or something, they've got 'em, they give 'em out at the start, so….Thinking of that before the start. Little bit more coffee. Then I'm ready."

"Some sandwiches, and *patisserie* mainly, you know, cakes…"

The racer's input displayed. On trays in the foreground the chef and his staff have laid out what was eaten by one racer for one day: some six or eight square meals. For comparison, my own day's food is on a tray behind the racer's.

"Des quiches, des tartelettes, du pain…"

"Some fruit, little tarts…"

"Maybe some fruit afterwards…"

"Nuts and raisins mixed together with milk…"

"Fruit, stuff that's easy on your stomach. Wait a couple of hours, then we have dinner."

And about dinner:

"Pasta, rice, chicken. Maybe steak…"

"Lots of pasta, rice, some steak…"

"You can use up an awful lot of calories, so…you have to put them back in."

We took dinner with one of the well-known teams of the Tour de France, and enjoyed the meal and the conversation, especially the data and technically sophisticated comment of their celebrated director, Cyrille Guimard, and his staff. Then we went to the hotel kitchen to share the experience of Chef Jean, the man who knew best what the team had eaten that day. He and his staff laid out for us on tray after tray the same dishes they had prepared and served that day to every member of the team. For comparison, I told him what I had eaten (it was also his cooking) that day, for I had kept careful notes, and we spread that on a tray as well.

We can inspect the fuel for one day for one man racing in the Tour de France. It is familiar, attractive food. We all recognize it as good French food, well prepared, dish after dish. There is nothing surprising about that; nothing is unfamiliar or mysterious, no wonder food.

What *is* different is its quantity. That is surprising: an enormous breakfast, a sturdy packed lunch served and eaten on the road without stopping the race, the big dinner that we all shared, and a couple of substantial snacks that had been served both before and after the six midday hours of the race.

My menu was the three square meals of a tourist enjoying France (plus a little afternoon snack, and an evening bottle of beer). But simply by looking at the loaded trays of food it became obvious that what I ate was very much less than the array of a racer's diet.

I do not work at the rate of the racers of the Tour de France, and the difference is apparent. We measured their impressive intake with some care.

TO MEASURE, ONE MUST HAVE A UNIT. I wanted to use a homely, evocative unit, quite informal, concrete, and without a Latin name.

My choice was an ordinary jelly doughnut. Our unit of food energy will be the amount of energy released by one particular change, the combustion of a simple jelly doughnut in air. Call it the JD. We listed and roughly judged the

A RACER'S MENU

A physician for the best-known of American over-the-road bicycle races asked a U.S. rider to list on our behalf "a typical day's meals on the road." Here is his authentic list.

Breakfast:

large glass of orange juice
2 large glasses of milk
Double stack of pancakes with lots of butter and syrup
3 strips of bacon

Snack while riding:

2 Snickers bars
1 banana
4 pint bottles of slightly flat Coke

After the race:

1½ cans of Coke
1 can beer
1½ granola bars

Lunch:

barbecued chicken
2 large servings of potato salad
4 or 5 pieces of bread and butter
2 glasses of milk
2 pieces of pie

Later:

1 apple
1 candy bar

Supper:

a large piece of roast beef
2 baked potatoes with butter
3 slices of bread and butter
salad
2 servings of pudding

Later:

1 candy bar
1 malt shake
1 hamburger

But he woke during the night a little hungry, and sought out this snack at 2:00 A.M.:
⅔ quart of ice cream
4–6 Oreo cookies

The list is the actual day's menu of a hard-working cyclist who ate well in the American idiom. The racer's high energy input is international; this day at table amounts to six or eight American meals, more or less what we found for the racers of the Tour de France.

Thirty-two jelly doughnuts and their fire.

size of every serving of food taken all day long by a racer of the Tour. We then used the published tables of the nutritionists (all dieters will know of them in one form or another) to compute the result in standard terms, reckoning foodstuff by foodstuff and weight by weight just how much food energy a racer takes in each day. We matched that result in our own JD units.

So unit by unit I collected first one jelly doughnut, then another—plenty of doughnuts were available! When I had assembled about a dozen, that matched my own everyday diet, and roughly yours. A dozen JD is a reasonable estimate of the usual diet of an American male who is not too large or too small or very active. (Of course, the variations are wide.)

But that would not suffice for the racers by a great deal. Those men with their remarkable physical performance go far beyond a diet of a dozen JD. They work remarkably hard, they manage a big input; to match it we needed to assemble many more doughnuts.

The reader is fairly to be warned: I am not making any nutritional recommendation whatever. Neither you nor I nor the racers can subsist long on jelly doughnuts, any more than the dietician imagines that you eat calories instead of carrots, beef, bread, apples. The human diet is quite otherwise; food serves more than one function for the body, and it must be varied.

But as a simple comparison unit for food energy alone jelly doughnuts work beautifully.

I piled up a heap of doughnuts that matched the daily food input of a Tour de France racer; it is some 30 or 32 JD. (I could not and I need not be more precise than that.) That pile measures the energy income of one of those athletes each day.

We can convert a doughnut, or many of them, by burning it in air to generate heat, light, motion. A bonfire of doughnuts makes the racer's input vivid. I could feel the heat, see the flame rising; the smoke and the sparks drifted upward. We burned the 32-JD equivalent of the racer's diet. We had started our small bonfire with a little kindling, as most people would do. First we ignited some cut-up doughnuts out of the 32; that was fair. But we used a little paper and kerosene as well. What they added to the fire can be neglected; the starting fuel burned out in a very short time.

When you buy a jelly doughnut from the baker, you can eat it at once. But to burn it easily in an open fire, it must be thoroughly oven-dried, or, like green wood and for the same reason, it makes a poor fire. The fire was swift and the flames very hot. It was over in half an hour or so. Fire and the human body do not work in the same way. Within the body of the racer, the consumption of fuel by the much subtler chemical processes within the living cells takes all day, and there is no hot flame at all. Does that make a difference to the energy account?

We have known for a hundred years from theory and experiment that as long as you begin an energy release at the same starting point and take it on to the same end point, the net energy release is just the same, no matter what processes you follow from start to end. Here we have a common starting point, doughnuts and plenty of fresh air, and about the same end point, the gaseous products of combustion, which are surprisingly similar both for the body and for the fire. It does not matter whether that release is swift or slow, or by what means the conversion is carried on, by way of the complex metabolism or by way of the simpler hotter flames. (The small amount of fuel value left behind in the unconsumed wastes in both processes makes little difference, and could of course be accounted for in precise work.)

To count up the thirty-two jelly doughnuts that equaled the energy of the racer's diet, those six or eight ordinary meals a day, we had to translate the foods we saw them eat by using the published tables of the calorie value of foods. But how did the experts find those values? One way is to do just what we did, by a fire. They burned small samples of each food— doughnut, butter, sugar, flour, anything—tak-

MEASURES OF CHANGE, UNITS OF ENERGY

Changes by combustion in air—fire, body or engine:
   1 jelly doughnut (1 JD) equals about 250 calories
   1 jelly doughnut   equals   1 million joules
   1 gallon of gas   equals   140 JD

Other changes:

1 kilowatt-hour electrical energy, *a few cents worth at retail*   3.6 JD

1 pint of water evaporated near room temperature takes off 2.4 JD

1 pint of water heated from room temperature to scalding hot   0.1 JD

1 ton weight falling 300 feet yields 1 JD

1 ordinary toaster running for 1 hour uses 3 or 4 JD

   One jelly doughnut (1 JD) is 250 nutritional calories. A nutritional calorie is the energy needed to heat one kilogram of water through one degree Celsius. This usage is being replaced by the term kilocalorie.
   All of our numbers are rounded off.

ing care to confine the fire and to measure all the heat that came out. It is not easy in practice, but simple in concept.

This is how they do it. They use a small strong steel cylinder with a cavity inside, but very well sealed by a heavy screw cap with fittings for oxygen and electrical inputs. They place a little weighed sample of any food substance within—always dried, usually powdered. Seal all tightly, and add plenty of oxygen gas under pressure.

Then they can ignite the contents with an electric spark. A rapid reaction occurs, a little confined explosion. (I like to think of it as a doughnut bomb!) Nothing is allowed to fly away; all the products are caught inside that strong steel cylinder. But the reaction, like our fire, gives out heat. The heat slowly flows out into a large, well-stirred insulated container of water within which the cylinder is immersed. All you need do then is measure the temperature rise of that water: calories. Finally, the experimenter will open the cylinder to make sure that combustion was complete.

Accurate calorie values are compiled for all the foods you can buy. A standard calorie for the nutritionist is simply a unit of heat energy, the amount of heat that will warm up one kilogram of water through one degree Centigrade. Experimenters measure the heat released directly for a known sample and use the weight to estimate the calorie content of any portion of food. That goes on the label or the menu.

Foods are so varied that nowadays it is often easier to measure the calorie release of their several ingredients, and total it up that way.

The typical Tour racer enjoys a daily diet of about 7,500 or 8,000 calories a day, two and a half or even three times the usual adult diet.

WE HAVE GOOD EVIDENCE FOR THE RIDER'S energy input; it made a hot fire we could feel and see. We estimated it more quantitatively, too, as some thirty or thirty-two JD per day. But to complete the account we must pursue another question: how does the cyclist spend the fuel value that he takes in, how much for what ends?

We should note that he does not store up material to gain weight, or burn reserves to lose it. That is certainly conceivable, but in fact over all three weeks of the Tour a professional racer is seldom found to gain or lose more than a couple of pounds, unimportant compared to all those meals he eats, all the fluid he drinks, all the oxygen he breathes. His weight goes up and down a few pounds each day, but no more over all three weeks. As an estimate he takes in about 600 or 700 pounds over the Tour; four-fifths or more of that intake is water. He lives scrupulously on his energy income.

Now it is our task to look one by one at the ways the cyclist changes the world as he moves along, and to estimate how much of what he spends goes into each of them.

The rider's muscles are powered by his

In the wind tunnel a stationary cyclist feels the sharp relative wind as the engineers trace the airflow by a local injection of smoke.

The racers following each other are taking advantage of the air moved aside by those ahead.

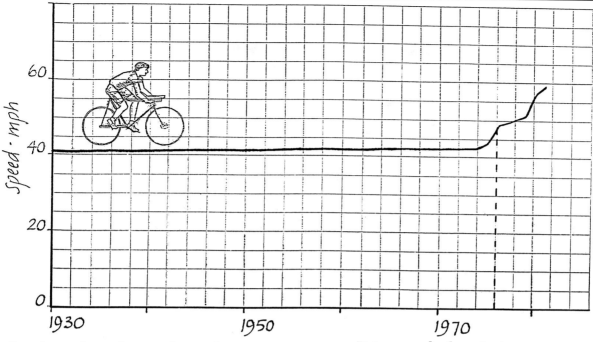

Speed record over the years for a cyclist over a 200-meter distance after a flying start: steady slow improvement, followed by the revolution of streamlining.

abundant food and the urgent breaths he takes. But what changes does that fuel make in the world? We first look at the mechanism, the bicycle. Everyone knows the pedals, the gears, the chain, the wheels....We expect a little noise, some heat or friction, as the legs and the bike move. They are there. That is real. Even more important, as the slender, hard, sewn-up tires go down the road they are steadily flexed as the area of contact between tire and road is loaded and unloaded steadily during the roll. Many people have felt the heat in a tire that has just stopped moving. All that is true...and yet the history of speed cycling demonstrates plainly that very little of what the racer uses is spent in the efficient mechanism of the bicycle.

Look at the graphic report of cycling timed officially by the Union Cycliste Internationale. The curve shows the speed record in the last fifty years for a distance of 200 meters with a flying start, a condition for very high speed. From 1930 to 1976 or so there is a slow increase in speed; the tires are better, the bikes are better, the training and technique of the riders slowly improves over the years.

It is a gradual evolution: incremental small improvements are possible. But all at once the best speed posted shoots up, in a few years increasing four or five times as much as all the gain accumulated over forty years of previous effort.

The engineers had got at last to the heart of the matter. What they found was *streamlining*. Even a glance at the new record-holding bikes makes the matter obvious. The tires are not better, the mechanism of the bike is not better, the cyclists remain as they were—but the air slips much more easily past the bike and rider. It must be that the air drag had set the limit. It is pushing aside the unseen air, half a ton or a ton of air shoved aside every minute, that uses up most of the racer's effort at motion.

The clearest demonstration is this. You watch a new streamliner and a standard racing bike running side by side, keeping pace. At a signal both riders stop pedaling and begin to coast. Swiftly the streamliner draws well ahead as both of the cycles slow down.

IN THE NINETEENTH CENTURY IT WAS WELL established how much heat was associated with motion. The engineers could time bicycles as they coasted down measured grades at all ranges of speed. By checking the speed changes of cycle and rider they could relate the energy expenditure to speed. Mechanism and tire losses are more important at low speed, but air-drag effects easily dominate at racing speeds, as the streamliners' record-smashing proclaims without any calculation.

From such measurements the expenditure of energy in air drag can be tabulated for every speed and each type of cycle. The result is clear in our case: the number of JD units the cyclist must spend moving both the bike and the air at Tour speeds is about eight or ten per racing day. Most of those are spent in moving air; not more than one or two of those JD go to roll the racing tires or to keep chain and gears going. That leaves twenty JD, or even a few more, unaccounted for. The account is not yet close to balance.

But never have we seen any energy account fail; it surely does not fail in the Tour de France. We must have missed some important channel into which the fuel energy flows. It is probably invisible, or we would see it. It is probably unintentional, or we would have thought about it. How do we balance the account?

A DIRECT AND DRAMATIC ANSWER CAME from another form of cycling: human-powered flight. In the spring of 1988 an ambitious team will attempt to send its specially designed pedal-powered aircraft along the route of the old artificer, Daedalus (may it not turn out to be the route of his imprudent young son, Icarus!). The plane and its pilot-engine, man or woman, must span about seventy miles over the waves from the tip of the Island of Crete to the nearest point on the Greek mainland.

Once before, such a gossamer aircraft spanned an arm of the sea under human power alone. The English Channel, a stretch of about twenty-five miles, was overflown in 1979. That pilot-engine was all but exhausted by three hours of heroic effort. The Daedalus route will require such effort for more than four hours.

The Daedalus development team, with the help of a leading candidate pilot, Lois McCallin, a 120-pound athlete of proven high output and endurance, approached the problem with care. They set up tests of cycling in a physiological laboratory at Yale, arranging that the stationary cycle used would put about the right demand for energy on the pedaler. In the earliest tests, McCallin could not long continue; she gave up well before one hour. The reason was plain; her body-core temperature, steadily monitored, had risen toward fever. No cooling flow of air had been provided. They knew at once what was needed: powerful fans.

We all understand it out of our own experience. Hard exercise is warming; you heat up and even sweat. In any brief bout of physical action, the body can store heat, warming as it does so, and then lose it slowly later, when the

Human power in flowing air. The silvery streamlined fairing covers the knobby and complicated mechanism of the three-wheeler next to it, to lead the air smoothly past; that control over drag brings record road speed to the cyclist. Similar streamlined form—and a cunningly shaped wing—allows another cyclist to fly.

athlete cools off at leisure. A ten-second sprint needs no special cooling. But for a sizable output sustained over hours there must be a way provided to remove ample heat steadily from the body. The air that flows swiftly past the moving racer amounts to a strong, steady wind. It is that wind which takes away the cyclist's heat. In the lab the stationary cyclist was in almost still air; but on the road in the Tour de France the racer can count on a twenty- or twenty-five-mile-an-hour relative head wind, produced by the racer's own motion through the air. The wind streaming against face and chest is what carries off the waste heat.

The plastic streamlined enclosure that surrounds the pilot-engine of the Daedalus craft was redesigned. A large tube, its opening aimed forward into the airstream, will take in fresh, swift-moving air to spread its cooling currents over head and chest of the pilot within the enclosure. A second opening at the rear of the enclosure will discharge the warmed air out into the wake of the aircraft. A careful compromise has to be sought between the necessity of sufficient airstream to cool the hard-working engine and the increased air drag caused by any opening in the smoothly streamlined enclosure.

Suppose indeed it is this loss of heat that balances the energy account of the racer. How is it taken away, and can we really balance the books? An automobile engine will help the argument. It runs very hot; we all know that. For every energy unit yielded by the exploding gas it burns, a good engine can transform only 25 percent into wheel-turning output. Most of that heat goes off as waste into the airstream through the radiator, much less with the hot exhaust gases.

The auto engine is water-cooled, they say. But engine cooling is a two-stage affair. That cooling water flows only under the hood of the car itself, a flow that carries heat from the interior of the engine out to the finned copper tubes of the radiator, where the rush of the passing air takes it away. But the water is sealed in, circulated over and over internally; a modern car in good shape rarely needs added water. The engine may be water-cooled, but the radiator is air-cooled.

For every energy unit, every doughnut, in the racer's food, only 25 percent can be transformed into muscular output. That is like the engine. But in losing that waste heat the racer's body is more subtle than the hot-running engine. It is air that finally accepts the heat from the flowing blood, all right, but another step can intervene. Normally it is the moving air of gentle ventilation that cools the body; asleep in cool air we even slow the flow of body heat out to the air by blankets.

But there is another item in the racer's intake we have not yet entered into any account. He eats well, yes, but he drinks water

TWO ENGINES AT WORK

	CYCLIST	COMPACT AUTO
Weight overall	170 lbs	1 ton
Speed	20 mph	55 mph
Total output	about 1 kilowatt	about 55 kilowatts
Lost in mechanism and tires	5%	8%
Lost in air drag	20%	18%
Lost as heat	about 75%	about 75%

even more vigorously. You can see that on the road and off. A typical racer on a typical summer day takes in four gallons of fluid, more or less. That is some thirty pounds, much more weight than all the food he eats. Weight balance alone shows that he must get rid of it effectively.

The body normally excretes three or four pounds of liquid water a day, maybe more. The rest of the water must go off invisibly as vapor, in the breath or from the skin. But that implies a considerable change in the world. To turn the cool liquid water that the racer freely drinks into water vapor that invisibly leaves his skin, or is sent out in his exhaled breath, takes heat energy, about one JD unit per pound of water vaporized.

A teakettle spouts out water vapor furiously as it boils. It gets no hotter; most of the energy that flows in from the hot flame beneath goes right out of the kettle spout in the water vapor into which the liquid from the tap has been transformed. That kind of evaporative cooling is much more effective than simple dry airflow. It is twenty times more economical in water used per unit of energy to evaporate it away than to shed warm water as liquid beads of sweat. Sweating is in fact a kind of failure of the cooling system of the body, a sign that cooling has reached some local limit.

NOW OUR DOUGHNUT ACCOUNT BALANCES neatly. The 8 or 10 JD of energy spent chiefly on pushing aside air must be augmented by 20 JD more lost in heat. Can all that water be accounted for? Quite well; the heat is carried off mostly by changing drinking water into water vapor (there is always some dry-air cooling as well). The loss is about one JD for each pound of liquid water that is changed to a pound of water vapor. The two insistent accounts then balance—we have rounded off all our imprecise numbers—both for weight and for energy.

The inquiry has shown some surprises. It has given the insight we hoped for. The fast-moving air past the racer is certainly his enemy, for it is his main impediment. The streamliners proved that. But it is also his best friend; the troubles on the stationary bicycle showed that. It takes a strong wind—I am not sure just how fast—to carry away the water vapor of cooling swiftly enough so that most of the water he takes in is not inefficiently lost as sweat. That would result in the overheating he must avoid. (No human racer can drink twenty times the four gallons offered in the Tour.) Those streamliners will be able to set endurance records, and not just sprints, when their designers, like the engineers of the Daedalus project, arrange for plenty of cooling air to enter the enclosure,

The cyclist takes off. The prototype of the Daedalus aircraft is shown on a takeoff run. The hundred-foot wingspan of this elegant flier is about the same as that of two-jet passenger airliners, yet the entire pedal-powered plane weighs under ninety pounds without its pilot-engine. No cooling had yet been provided for; the next version would admit cooling air. The design and construction of this aircraft were chiefly the work of MIT students and faculty.

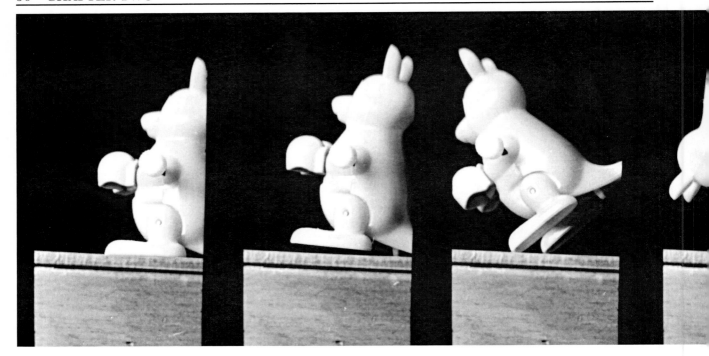

even at the expense of streamlined perfection.

Very few engineering products work as well as the human engine. Even so, most of its fuel is spent as heat, and that heat must be carried away somehow or the body will suffer the maladies of fever. The Tour racers owe their powerful performance to the splendid cooling provided by the swift self-made airstream whizzing past at twenty and twenty-five miles an hour. A distance runner moves more slowly, feels less wind, and could not lose the cyclist's heat output on a summer's day.

The wider lesson is also important. Whenever the energy account does not quite balance it is better to look for a missing channel into which the energy is flowing than to imagine you have discovered an exception to the universal rule to which all experience so far has led us. The energy account will balance once you strike it properly. Every organism, every vehicle, and most other complicated systems can be well understood only after working out their weight and energy balance, two deep constancies in nature.

What the Twentieth Century Found

THINK ONCE MORE OF THE SMOKE-FILLED box on the old farm scales. The result we found from our own Rube Goldberg variations was a straightforward one. The weight of the box stayed constant in spite of all the changes within because the box walls were solid, sealed, impervious. No gas, no flying bits, no substance, no material, could leak out.

But is it correct to say that nothing at all leaked out? Light certainly leaked out, for everyone there saw flame, sparkle, and flash. Sound certainly leaked out, for we could hear the squeak of the toys and the loud reports of the firecrackers. The radio kept working, so it must have received some signal through the walls. Heat, I can assure you from my place nearby, came out in plenty.

Are all those *nothings*?

Certainly not. They are all forms of energy, themselves accountable under the nineteenth-century discovery of energy balance, just as we used heat in the Tour account. But since they

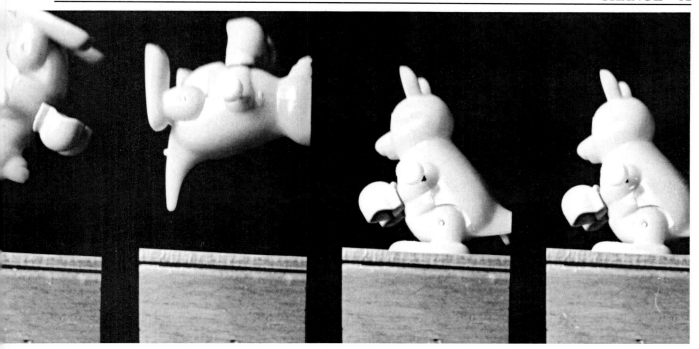

are not substances, not material, since they contain no chemical element, no atoms, at all, they have no weight either! So it is no wonder that in spite of the manifest leakage of heat and light and sound the weight balanced.

Who could ask for more?...It was Albert Einstein who could and did ask for more, and he answered with more, too, in the year 1907, a superb physicist at twenty-eight. Others had given hints, but his was the first persuasive argument that proposed the result we all know now in words.

$E = mc^2$. That formula (along with H_2O) is one everybody recognizes, if not perhaps so many know what it means. What it means is wonderful, and even easier than the much-quoted expression.

Energy in every form, said Einstein, has its proportionate mass. Mass is only the physicist's word for anything that has the property of weight. Energy and mass correspond and are equal.

We can and we should write the expression, quite correctly, in a simpler form:

E, energy, equals M, mass.

(That c-squared is only a constant quantity that reconciles between one kind of unit for mass and another for energy. Choose your units for simplicity and the c-squared drops out.)

$E = M$: that's what it says. Every change in energy requires an equal change in mass, whether loss or gain. The two go together. When we perceive the energy leaking out of the transparent box as heat and light and sound, we have to accept that mass leaked out of the box as well, and so its weight changed, its weight decreased. Yet we found no change, to a part in one thousand. The chemists did much better even than that.

There is only one way out of that trap. The weight loss is real just as Einstein said. But it is a loss so small in amount that it cannot yet be measured even on the best balances.

Take a playful example, a windup toy. When I wind it a spring within, at first relaxed, becomes tautened and distorted by the force and motion of my fingers. That is the form in

THE WEIGHT OF ENERGY

A windup toy unwinds and lightens by one billionth the weight of the dot of an *i*.

A car burns a tankful of gas and plenty of air; the total weight lightens by 10 dots' weight.

A thermonuclear bomb explodes and the total weight is reduced by a quarter of an ounce.

The sun shines for 1 second and loses the weight of 2 dozen ocean liners.

Of course, all the weight appears elsewhere.

which I add energy so that the toy can go through its tricks. But that energy, which I might measure by letting the spring unwind to warm a tiny amount of water, must imply that the spring has actually gotten heavier. It has gained in mass as I wound it, as of course I have lost a little mass by the act of winding it.

The spring is made of steel; certainly no steel flowed from my fingers into the taut spring. The stressed, taut spring is simply more massive—heavier—than the relaxed state of one and the same piece of metal. Mass has come in because the energy of that stress has mass.

Now let the toy go through its paces. As it hops the spring is unwinding, relaxing, less distorted, losing energy and therefore growing lighter, actually losing mass. Of course, the mass cannot disappear—it goes somewhere. It went out with the energy, the sound of the toy's hops, the minute wear of the table surface. Those places in the environment have now gained all that tiny bit of mass which was once stored in the spring from my fingers. That's the way the proposition must work.

But how much is that change of weight?

The rules that Einstein laid down enable us to estimate how much mass is taken up by the little spring and given out again to the environment. It amounts roughly to a part in a billion out of the weight of a pencil dot over the *i* in my name.

There simply is no way to use the ordinary instruments of weighing to check up on this wonderful result of Einstein. But we believe it to be true, and it has been verified to high accuracy many times. The valid tests come only from the atomic and subatomic laboratories where they have long known ways for measuring the mass without the use of gravity.

Those investigators do not work with bulk matter, with any solid or liquid bits, however small. They turn their samples into very dilute gas, vaporizing them into streams of individual atomic or subatomic particles enclosed in a vacuum, where they are free from the stray material of the world outside. They deflect these beams, not by gravity, but by electrical and magnetic forces. Indeed, that is what happens to the electron beam within every TV picture tube, deflected artfully to paint point by point the picture that glows on the screen afresh many times each second. With similar elegant control they can measure the orbits of the streaming particles in any beam under a known force, and so find their relative weights with precision. They use the atoms one by one, and not in bulk. Finally, they "weigh" the initial and final partners of nuclear reactions, and compare the mass difference with the measured energy released. Agreement was first adequately tested in experiments about 1935; by now the result is past doubt, with an accuracy of parts in a million.

EINSTEIN ON THE SCALES

THERE IS A HOLE IN THE NET IN which the physicists caught and tamed the idea of energy in the twentieth century. No doubt exists now about Einstein's equivalence of mass and energy. The relation has been checked very well indeed, though never by direct weighing on the scales. The strong theoretical basis was convincing from the first, but now experiment confirms it as well. The mass measurements all come from the use in one way or another of electrical and magnetic forces on streams of individual particles moving in the vacuum. For the mass measures the response of an object to forces that push or pull, and whether it is gravitational or electrical makes no difference. There is nothing esoteric now about these techniques; they are used in a great many labs. The chief instruments are called mass spectrometers of one kind or another, and particle accelerators offer closely related evidence. But none of that is the familiar operation of weighing bulk matter. That should work, too.

I wanted to exhibit the direct and delightful implication of Einstein's famous rule: put some well-sealed but internally changing system on the pan of a balance, and wait. As that box emits heat, light, even sound, it will lose weight as you watch. This is the same kind of experiment Tom Tompkins built for us in a Lucite box, but now pushed to very much greater delicacy.

But I couldn't do it, even with some help from my physicist friends. I worried them with questions at teatime. They found it interesting to ask among their own friends at other teatimes. It must be possible! After a while proposals begin to come in from people far away whom I did not even know, but who had heard about our problem. None seemed workable. I think it can be done, but it would be a long job, unattractive because we are all so sure that it must work.

Why is it hard? Consider how to do it. If

The glowing sphere is about as powerful a radioactive source as can be handled. Its active core is six ounces of plutonium dioxide (the isotope Pu^{238}), clad in a thin iridium metal sealing shell inside a strong graphite mantle. It puts out 100 watts of heat, as this glow witnesses, and it will keep it up for decades.

you use a big change, with lots of heat coming out, it ought to be easy. But recall, the system has to be sealed. A jet engine puts out plenty of heat but it draws in and puts out very much more air than the tiny weight of energy that leaves in heat and motion. So you would require the engine, its fuel tank, its air, and everything else sealed up tightly in a room. An explosion is even worse. Nobody wants to weigh something like that. Wires to bring in power will not work; they are bringing in weight as well.

You need something self-contained. A wind-up toy is fine, or a chemical reaction inside a container, or a battery that would steadily discharge by lighting a lamp. Those might be held on a sensitive balance. But they won't do. The change in weight during a chemical reaction is a few parts in ten billion of the total weight of the reactants, and it is hard to build a balance with such fine discrimination, especially if it is going to be heated or jiggled about during the process. The fine ultramicrobalance we used fell short by a thousandfold. And what about the container? You won't find it easy to build a sealed system without adding a good deal of weight to contain the active ingredients. The thirty-two doughnuts do not seem hard to weigh, but the oxygen to burn them would come in a heavy steel cylinder or two. The best little batteries, used in digital watches, are sealed and easy to weigh as they discharge. But the change in their weight during discharge is just too small.

So the last hope is a radioactive source, small but fierce, well sealed so that only heat and light and radiation leak out. Now the fractional change in weight can be appreciable, easily up to a few parts per hundred thousand, and perhaps well more. How much energy leak can we provide? If we prepare to watch the weight dwindle, say over a day or two, we can tolerate the power of a light bulb—maybe—on the delicate balance. (We might feel it easy to seal up a midget firecracker, but a grenade is daunting.)

I found the people who had elegantly prepared the hottest safely-sealed radioactive heat source anywhere. It was one of a set made to place on the moon to power appa-ratus left behind there. The source looked something like a graphite tennis ball (full of pure radioactive plutonium!), but it can keep itself glowing orange-hot in a vacuum for decades. No microbalance would enjoy that, and yet the heavy little object would lose under a microgram a week, the weight of heat and light and radiation given off.

Would not its surface dust off a little, or corrode that much? We might need to place it in an evacuated container that would weigh plenty. Would the gas that built up inside the seal as the plutonium gradually released helium atoms leak out? In fact, its knowing builders had arranged just such a slow leak. The otherwise tightly sealed sphere was ventilated at its two poles to relieve any pressure by two little spongy inserts of sintered metal. They keep in radio-active material, but allow the helium-decay gas to leak out. The leak would have to be sealed while we used the somewhat unap-proachable sphere. The weight of helium produced in any interval of time is a thousand times the weight of the energy lost.

I gave up. Maybe somewhere there is the right energy source, easy to weigh, easy to handle, powerful enough to see the change in weight within a plausible time. Or maybe there is a quite different way to go about the whole thing?

Let me know when you find one...

THERE IS A GAP HERE FOR THE PHYSICIST. I tried hard to find some context in which we could carry out the test of the fundamental Einstein relationship by weighing bulk matter. In the end, there was nothing satisfactory. The gap really means that we have two definitions of weight, one with bulk matter and the other using atoms one by one. The relationship is not yet known with sufficient precision to allow a check on Einstein by the ordinary means of weighing. It does not seem beyond present possibilities, only difficult, taxing, expensive in time and money. Someday it will be done. But no one expects a surprise; the particle-at-a-time agreement is really excellent and the result is well confirmed by many distinct instrumental techniques. Still, a gap remains; here the fabric of science is not quite seamless.

That is why most people have come to think—quite wrongly—that $E = M$ is a statement true only of atomic and subatomic processes. Not at all; it works in all energy transfers. But only in the atomic labs has it been directly verified, and only in nuclear changes is energy really an appreciable fraction of the total mass involved.

$E = M$ does hold in every process of energy transfer we have ever seen. Throughout them all, the joint account, including both energy and mass, remains in good balance to the accuracy of the experiment. In ordinary affairs the two can also be regarded as distinct accounts, for the weight balances and the energy balances as well, each on its own account, since their values are so different.

Energy is conserved, and mass is conserved. The two remain useful even as they merge into one. Call it energy, call it mass as you will; it acts as one quantity, constant and unchanging throughout every change: the fast and the slow, the large and the small, the profound and the superficial.

Pocket globe of 1780. This is the mapmaker's image of the world, the round earth held within the ordered blue sphere of the stars.

3

MAPPING

THE GLOBE OF THE EARTH IS MORE THAN IT once was, when it could be seen only as a carefully produced colorful model in the library. Its nature and size have been known well for a long time, even stirring the imagination of the poets. But now we see it for ourselves easily and compellingly in photographic images taken by satellite, while we watch nightly the march of the weather across our whole continent.

The mark of our species in those whole-earth images is clear only by implication, in the camera's chosen position. Photos that show the planet from an American-made orbiter over the equator, its motion in space timed to follow the spinning earth below, usually pose equatorial South America in the foreground, while the images made by a similar satellite working for the Europeans center on the meridian of Greenwich, and prominently display the western portion of the African continent.

But on the surface of the earth itself there is very little sign of human activity in all those sunlit images. Once the stars come out the pictures are transformed. Creatures of the sunlight, we have found the artifice to light our cities, towns, and roads even after earthspin

has carried them into the shadow of night. A couple of unusual cloud-mapping satellites operated by the Department of Defense are able to image light as faint as moonlight from a height of about 500 miles. They scan the earth constantly. Their images show the earth at night almost entirely as it is picked out in the lights and fires of man.

The great metropolitan centers stand out by their own glowing lights as plainly as on the best maps. Coasts and rivers and even highways and railroads can be traced by the gleaming beads of the towns that are strung along them. In season we can see the brush fires set to clear the fields over whole countrysides in regions of Asia and Africa. Cloud-covered areas remain pitch dark, and so also the broad, unlighted oceans. A single house or a single ship cannot be seen.

The changing images even present a few riddles, some with surprising solutions. In the empty center of the dark Sea of Japan there was a scatter of light that rivaled the lights of nearby Tokyo, the biggest city on earth. It turned out that on this particular clear night the squid were plentiful, and drew there some two thousand vessels of the big Japanese fish-

THE EARTH

By night the video images from a special scanning satellite shows mostly the fires and lighted places of human beings. The cities of the eastern United States are ablaze with light; a little study would trace out rivers, highways, railroads by the light of the towns strung along them like beads. Western Europe shows London and Paris, the Rhine cities, Madrid, Milano, Rome.

The geography of night and day. Most satellite images are taken by daylight, like this one made by a stationary satellite of the European Space Agency. It orbits so that it appears to hover high over a point off the coast of West Africa, on the equator straight

FROM SPACE

south of Greenwich. It can claim the desirable conventional address of zero degrees latitude, zero degrees longitude. The continents show landforms plainly, though with little sign of humankind.

A riddle of night and day. By day the unpopulated Sea of Japan is blandly cloud-streaked. Then once a night city glowed there to rival Tokyo. The big squid-fishing fleet carries bright lights to bait their prey, and they had schooled there along with the squid.

Before the satellites. This map was published about a century ago. It displays a whole continent splendidly, though it was made on the ground alone, without help from any overview. How?

ing fleet. Those squid boats are equipped with scores of lights as bright as any streetlights, which they dangle over the water to attract the squid. Their merged light resembles a night city that has drifted to sea.

Those photographs are novelties of our time. They were all made by machines in orbit. From that high vantage point, the lens can easily catch a big scene of earth, a substantial portion of the great globe, in something not far from a snapshot, as the world changes in time.

But ours is not the first age to understand in detail the form of the earth on which we dwell. There are plenty of old books that are full, to be sure not of photographs of earth, but of maps. These books go back fifty and one-hundred years, yet their maps are impressive. Those maps were made by people who had no orbiting machines. They hardly had useful aircraft or balloons. The occasional mountaintop might offer them a distant view, but for a hundred miles or two at most. Yet they mapped the world as a whole, completely, accurately, and handsomely. Across the whole face of Europe, for example, all the cities are located in their familiar positions, and the boot of Italy thrusts just as we expect into the Mediterranean. None of the satellites brought real surprises to the world map.

How could they have done it back in the days when they were restricted to crawling like ants over the globe? That is the story I want to pursue, not only for itself, but as well because it is a kind of parable for the growth of understanding in many branches of science that lie far from the work of the geographers and the navigators.

Tom Jefferson's Garden

WE BEGAN ON THE GREEN HILLTOP AT Monticello, overlooking Charlottesville, where Virginian Thomas Jefferson—surveyor, architect, and Founding Father—laid out and built his model estate at the end of the Revolutionary War. (I am sure he would have enjoyed as much as I did the way many Virginian visitors we met there alluded to their famous old neighbor as Tom Jefferson.)

WE CAME TO BEGIN OUR MAP OF THE HOUSE and grounds by an enduring old technique, one still much in use though simpler than the standard of Jefferson's time. (Fred Wiseman, our experienced surveyor, is an archeologist who uses this same scheme to plot his excavations.) The essential framework of the map is drawn line by line as the work is done, pencil on paper. The paper is pinned to a flat board, called the plane table, usually held on a tripod so that we can set it up anywhere. Each sighting is made simply by looking along a triangular ruler laid on the board to the point we want to map. A pencil line is drawn down the edge while the ruler is correctly aimed.

Once we had drawn a set of rays to all the

Jefferson's grandchildren in the garden, represented in a watercolor of 1826 which shows the main house and the low walks that connect it to the small structures at the two ends of the U.

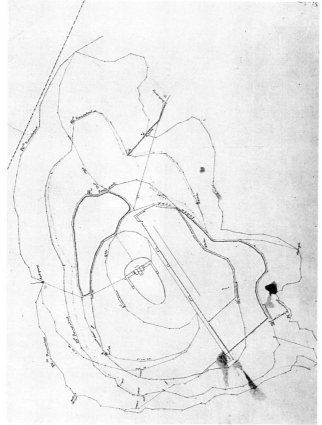

Our mapping party small in the big garden, as seen from a helicopter.

Monticello mapped by Surveyor Jefferson's grandson in the early 1800s

PLANE TABLING

The plane table party sets up.

Leveling the board. A cup of water would work, too.

The surveyor sights to a landmark.

One line is ruled. The triangular ruler serves both for sighting and for marking.

What he sees: a corner of the walks.

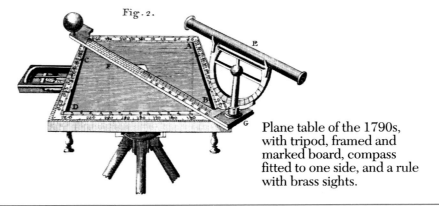

Fig. 2.

Plane table of the 1790s, with tripod, framed and marked board, compass fitted to one side, and a rule with brass sights.

places we want to fix, we prepared to move our tripod and our point of view. To do that, we chose another place on the lawn, and one of us stood there to allow a sightline—a surveyor's baseline—to be drawn toward the marked point. We measured that baseline with a long tape. When we moved our equipment over to the new point, first we sighted carefully back from the new position to a marker left at the first station, and then we repeated the sightings on the house and its dependencies. The new rays cross the previous rays one by one; each intersection of a pair of rays fixes that point on the paper. Distances are all in proportion; measuring any one distance (by pacing or by a tape measure), say the seventy-seven-foot walk we took from one viewpoint to the next, is enough to determine the scale of the entire map. (A second such baseline would offer a welcome check.)

The table needs to be level at each position; for that we called upon gravity through a small hanging weight, or we might have centered a bubble level. Tilted tables would certainly distort the plan.

We located as many points as we had the patience to fix. Any remaining details on the map are filled in freehand, but a freehand now guided and controlled by the well-mapped points of the framework: easy when those are many, harder and less reliably when those are

few. Our map can be compared to an older one; the old Monticello map was more professional—and took much longer.

Almost every map that claims to present the form and size of a place has these two elements: a framework to control sizes and shapes, and then much added detail. Maps usually bear symbols that identify interesting

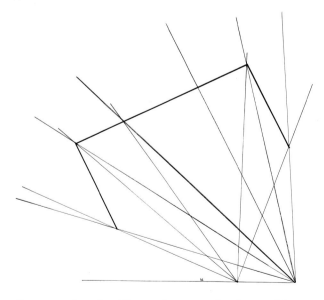

Our map that day. This is the start of a plane-table map of Monticello. Measuring the distance between our two sight points (or any other mapped length) would fix its scale. The right-hand corner of the big house was hidden from our second sight point behind the spreading copper beech.

locations not drawn to scale, like paths or roads. This simple map is a guide to a grand family.

How far can this method be extended? Look at a pair of sightlines, rays, that intersect to fix a location from two positions of view. Consider the triangle with its two sides the rays, and its base—baseline—the distance between the two sighting stations. Think of looking from the same two stations at locations farther and farther away. A few times farther seems not to make much trouble. But note that the triangle becomes very slender and pointed. The two rays are aimed close to the same direction; they must extend a long way before they intersect to fix the far-off point. Now a small error in direction will make a big difference to the judged location. Then imagine, if you can, going out a hundred times farther still. Then the two rays from your two viewpoints appear parallel to each other as far as you can tell; the remote intersection cannot be determined unless the rays are drawn with accuracy not to be reached by pencil lines on paper.

A FRONT-SEAT PASSENGER, I LOOKED OUT THE car window as we drove steadily down the straight road along the floor of a New England valley. The nearby row of roadside maples flew past so rapidly that I could not point out any one of them. Its direction changed too fast, moving from ahead and right to well behind in under a second. The farmhouse in the middle distance moved more slowly in direction, though I could hardly point at it long enough to catch my driver's attention. The more distant barn on the back lot slid slowly past the window; the hills went by at a leisurely pace; that distant mountain stayed at much the same bearing for a little while. But the truly remote moon will seem to keep pace along with me for minutes, visible always in the same place until we come to the first real turn in the road. The more distant something is the less its direction changes as we move along.

That same generality turns up in a different way, easy to recognize even when you are not sailing down the Vermont highways. Move your hand before your eyes and it quickly crosses the field of view. The birds fly by smartly, too, but not that fast. High clouds, even scudding along on a windy day, take a couple of minutes to change much in direction.

That familiar experience applies to the sightlines we drew at Monticello. Had we tried to include a distant peak a few miles away among the locations we sighted on the plane table, the little triangle we might have drawn for it would simply not have been reliable. The sightlines from the two stations would have gone out at best in the same direction, quite parallel, to form no triangle at all. With bad luck—some small shift of the pencil—the two rays might even have crossed absurdly enough

That which is farther away moves more slowly past. Watch the change in position of tree, barn, and hilltop as you drive by.

in the wrong direction, behind our heads while we were looking out to the mountain.

A plane table map can hardly fix locations more distant than a hundred times the biggest baseline you have stepped out on the ground. A map carefully made in the gardens of Monticello might suffice to locate the more distant hills around it, a few miles away. But with much bigger baselines and much improved accuracy of sighting—using not the naked eye and pencil lines but telescopes with cross hairs, mounted on elegantly graduated and calibrated angle-measuring circles—the same principles can be used to throw lines across country. The triangles might grow to the size of counties or even small states, as far as the vista from a mountaintop.

There are limits; a long sightline is always disturbed by the vagaries of a nonuniform atmosphere, even if everything else is well done. The decisive limit is encountered at the ocean shore. Just before the French Revolution sightlines were surveyed across the English Channel by night, to tie together the big triangulated map frameworks of Britain and France, but the Atlantic Ocean is not to be so bridged.

THE SURFACE OF THE EARTH—AND EVEN THE sea—is not flat; the bulge brings in a distortion to the triangles that grows as the map grows, one fully curable by mathematical sophistication after long-distance earth measurements.

Hills and mountains can be managed, too. There are two ways; if you look up at a hilltop or down to the valley from the plane table, but take care to measure the inclination of the sightline from the horizontal, you can work out any distant altitudes from the triangles that are now in the vertical plane. But it all depends on small angles, not easy to fix accurately. There is another way, leveled sighting in steps, very tedious. A vast leveled network of repeated large triangles across country was the basis of most mapping frameworks until the present electronic age, the plane table map turned into a continental giant by exquisite visual instrumentation and heavy calculations.

The sun also rises, and at day's end it sets again. We mark those terminal points, but sel-

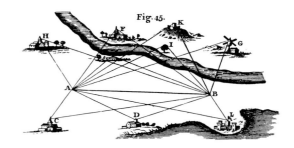

The triangles in the textbooks are never too sharp for good results. This is from an eighteenth-century text. Imagine for yourself how the tree across the river would appear to shift as you moved from point A to point B.

MAPPING THE HIGHEST MOUNTAIN

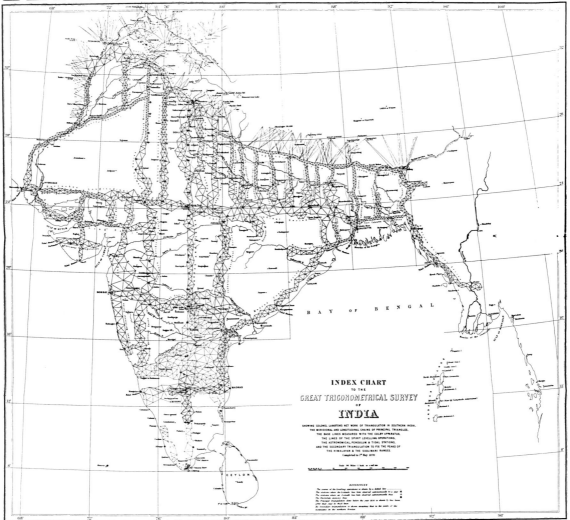

INDEX CHART
TO THE
GREAT TRIGONOMETRICAL SURVEY
OF
INDIA

SHOWING COLONEL LAMBTON'S NET WORK OF TRIANGULATION IN SOUTHERN INDIA,
THE MERIDIONAL AND LONGITUDINAL CHAINS OF PRINCIPAL TRIANGLES,
THE BASE LINES MEASURED WITH THE COLBY APPARATUS,
THE LINES OF THE SPIRIT LEVELLING OPERATIONS,
THE ASTRONOMICAL PENDULUM & TIDAL STATIONS,
AND THE SECONDARY TRIANGULATION TO FIX THE PEAKS OF
THE HIMALAYAN & THE SOOLIMANI RANGES.

THE SURVEY OF INDIA IS AN UNEXpected flower of the old mappers. Its network of big sightline triangles was meticulously stretched out across that subcontinent for generations. They used triangles in the vertical plane as well to find mountain altitudes at a distance, their transit telescopes sighting a hundred miles north and just a little up from the horizontal at the immense barrier of the distant snowy Himalayas. The stations used were towers built above the treetops on the damp hot plains along the border of forbidden Nepal. The many sight lines to the high mountains stand out from the northern border like an eyelash.

It was those triangles that found the highest mountain in the world. Mt. Everest was a mathematical discovery, made by mapmakers in the calculating offices of Calcutta. Close to the mountain even Everest is just the nearest towering high summit, and viewed from far off in the hills it is often dwarfed by peaks that are less high but much closer at hand. Only the calculated result of the triangles—enough of them to make sure, all sightlines carefully made in still air—can tell the truth and disclose the highest of peaks.

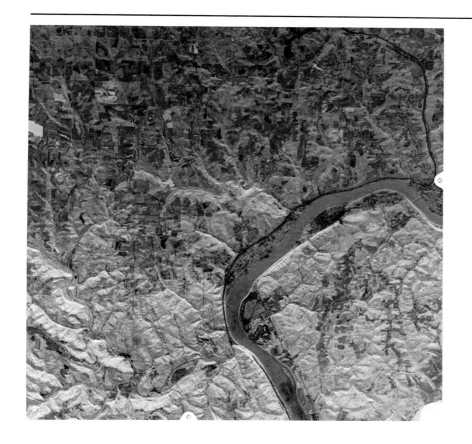

Aerial photographs of farmland
on both sides of the Ohio River
today. Checkerboard fields were
laid out when the settlers came
later than the surveyors. The
photographs show land use
within twenty miles of the
Point of Beginning, where the
surveys of the West started.
The great square grid, so noticeable on the flat plains, can be
seen to fall like a shadow even
over these hilly farms of eastern Ohio, too late to influence
the farmers five miles eastward
of them across the river.

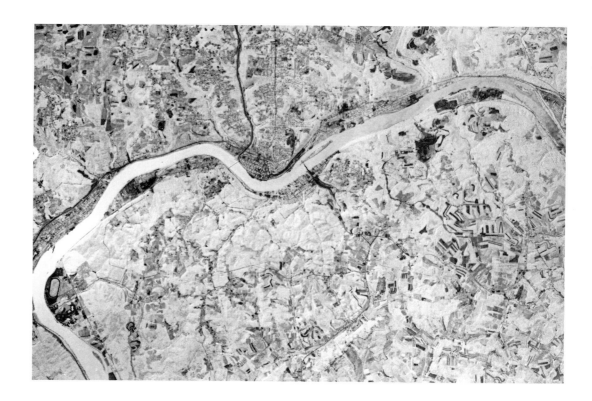

The bright constellation Orion, jewel of a winter's evening, moves across the sky. Several views show it as the hours pass; always the same in form, it sweeps past in the turning sky. You might note the direction where it stands at its highest.

dom watch its glaring arc across the sky from east to west. The starry sky executes a similar motion, more rigorously repetitive than the shifting arc of the seasonal sun. It is open to anybody to watch those glittering patterns wheel across the sky on a clear dark night; even a beginner is linked in awe by a few hours of such watching to a chain of human experience as long as any we know, an ancient source of the idea of natural order. The constellations to the south all rise, turn as one through the night, and come one by one each to its western setting. (Some stars circle to the north in arcs so small that they miss the horizon entirely, and fade only in daylight.) You, too, can in season see bright Orion, or some other star pattern, cross the sky. The motion is quite general; it includes any star you can see, always symmetrically, but in smaller or larger arcs. Every star will take the same path, its own path, night after night, even year after year, as long as you can see it, changing very little over a lifetime.

That is true everywhere and everywhen (some slow changes are visible over centuries). Star patterns remain the same worldwide. You may not always be able to see a given familiar constellation as the night or the seasons pass, nor as you travel the world, and certainly it is not always oriented the same way with reference to the local horizon. The old astronomers knew all that. It is interesting to compare the Orion we see now with the entirely convincing

Unchanging Orion was drawn much the same as we see it now by some watcher of the sky in far China 1500 years back.

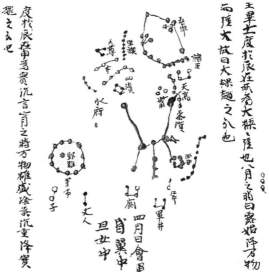

drawing of what a Chinese sky-watcher saw fifteen centuries ago. All star patterns, large and small, persist in time and remain rigidly unvaried in form, wherever and whenever you see them. The birds and the clouds move as they will, the sun, the moon, and the planets change and wander, but the stars sweep by rigidly as one over long periods of time. With that we seem to touch the cosmic order, the fixed stars.

How can that be? Surely you draw away from the stars in the sky north of you as you travel south. Even if they are lights hung on a crystal sphere, should not their patterns shrink at least a little as you move? The ancients drew a wonderful conclusion from all these circumstances. The "fixed stars," as they called them, must be quite far away, farther than any mountain, certainly farther than the moon. Sometimes the moon blocks one or another bright star for a while; it is hard to avoid the conclusion that the nearer moon has passed in front of the more distant star. Since the stars are so far away, they each appear in the same direction as viewed from any part of earth, the way a distant mountain appears from any part of the garden at Monticello.

How otherwise could their patterns stay so much the same over time and for differing positions on earth? Since stars do stand in a fixed direction from any point, they provide a firm celestial framework for maps of our whole earth.

The high point of each star's arc is unique. Follow the stately path of star after star; all will arc up above the horizon and then down (unless it is one whose circle is too small to rise and set across the horizon). If you note the high point on a number of arcs you will see that the sightlines to those points lie one above the other. Referred to position along the horizon, that common point is the south crossing, the same for every star. That is the very meaning of the direction south—and of course north is the opposite direction along the same line. (Viewed from the southern hemisphere, the account changes in an evident way, but the meaning remains the same.) East and west are nothing more than the directions at right angles to that north-south line. That the sun sets nightly at the western point is of course an exaggeration; the sun's setting is westerly, rarely true west, since it shifts day by day. The other traditional clues to the north and south direction—the moss on tree trunks, the magnetic compass, even the North Star or the Southern Cross—are only rough, though quick and practical, ways to approximate this defining, symmetrical direction.

We Measure the Earth

THERE IS SOMETHING AUDACIOUS ABOUT human beings, nominally some six feet high, who are able to gauge the form and the size of

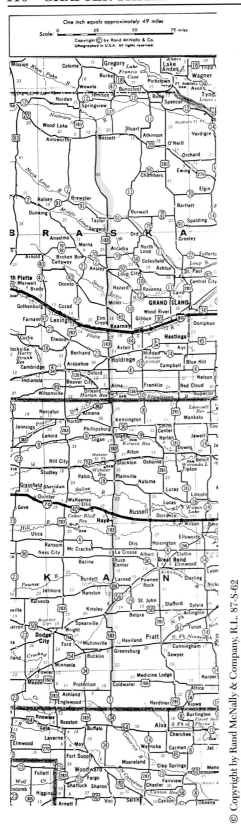

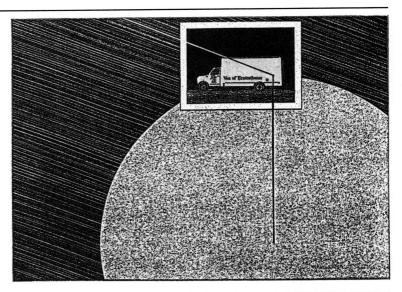

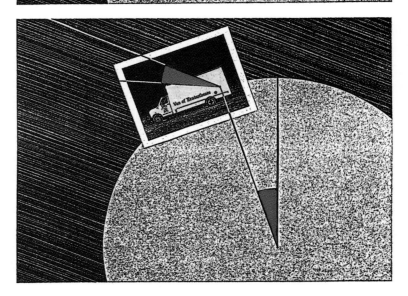

Straight south on U.S. 183. A strip from the standard highway map shows Route 183 in Nebraska and Kansas, the straightest long stretch of north-south road we could find in the United States. The distance we drove was about as far as from Boston to Baltimore, with only two gentle bends.

The sightline to Antares and the plumb line to the earth's center hold an angle between them. All the light from the remote star comes to earth along the same direction, as the textured sky reminds us. As the van goes along the curve of the round earth, that angle changes. But the change in angle is always the same on the van wall as it is at the center of the earth. (That change is colored red in both places.)

the planet we live on. It is an experience I had always hoped to try for myself, and not always to depend upon the atlases or the airplane pilots for that central fact in our traveling lives.

We came at last to the little prairie town that was our destination, and turned right at the highway crossroads into the main street. We drove a short distance until I caught sight of the high water tower, a sure landmark in these parts. It bore the town name written large: we were officially in Bassett, Nebraska, on U.S. Highway 183.

We tried the measurement starting a few miles north of Bassett, our principal instrument a fine yellow rental truck. It seemed right to name it the Van of Eratosthenes, for he was the first person ever to write about this scheme—and perhaps even to help carry it out—when he was librarian at the famous center of learning in Alexandria more than 2,200 years ago.

We had come to these wide prairies in the center of the North American continent because this very highway is the longest nearly straight stretch of north-south road in the United States. Continued far enough, U.S. 183 would connect the poles of the earth. We had chosen it at home one winter night by studying the highway maps; it is conspicuous there across the plains.

The road headed south, the proper direction. By parking the van aligned with the road

we aimed it straight southward. We had no sense that this was borrowing the work of others, because as soon as we saw the arcs of the stars we would confirm the direction for ourselves. U.S. 183 would save us time, and its straightness meant that the distance we rolled would not wander over the earth ball, but head along the right path from pole to pole. Hills and rivers are impediments to surveyors. Here in the plains they were minimal.

I practiced sighting stars along the vertical wall of the van. We had only to be patient until that bright star crossed at its highest point, right over the aligned van wall.

A bright star was essential to make the sighting practical. Our schedule demanded that we mount our expedition in spring, and bright stars that cross the south line in the dark hours are few. We chose the star Antares, bright red heart of the Scorpion, a star that would cross well past midnight of this cold spring night. Clear skies are essential for stargazers; we had waited a couple of days for them, until one late afternoon the clouds broke to allow the sun to brighten Bassett.

The night was not an agreeably clear one; we would remain anxious until the time of the sighting. But it seemed a chance worth taking, so we waited a few miles north of town on a local extension of the highway for Antares to pass the south point. Only a handful of cars passed while our picnic supper proceeded

The star Antares moves in the sky. The real sky is often graced (or marred?) by clouds and birds, unlike the diagrams. In this set of still photos the star Antares swings by on a cold night, crossing at its highest point right over the wall of the van, parked pointing south. In the last image the arc that Antares follows is drawn. The second bright object, just above the star, is slowly wandering Saturn, there in that part of the sky for many months.

in the dark fields next to the road. A bright wanderer lighted the sky; the planet Saturn would be near Antares in the south for the next weeks. But what I remember most vividly about the sightline we marked that spring night was the unexpected cold, frost on the stalks in the field, and a shivering wait for the star.

The sighting took cooperation. One of us looked up across a simple sight made of two closely spaced pegs we had mounted low at the rear of the van wall, while the other (on a ladder at the front of the van) moved a wooden rod toward and away along the van roof, until the rod blocked the glimpse of the star. When the star was hidden behind the rod the sighting was aligned as well as we could tell. The bright star passes slowly over the top of its arc, so it is not too hard to estimate such a fix. We marked our alignment by joining the best position of the rod to the sighting pegs by a straight run of black tape. The direction to Antares as seen from Bassett was then marked right there on the van; we would carry it south into Kansas. Our large and therefore fairly accurate sighting instrument was portable!

Here is one point more. We needed the direction of the vertical—that is, the line down to the center of the round earth. Gravity supplies that very easily; we dropped our plumb bob, and marked that direction, too, with tape right on the van. Once we had driven all day

Route 183, the long straight road.

south, we would park again, check the plumb line for vertical, and relevel the van if necessary by driving one or two of the wheels onto small pieces of plank we had brought along. Road surfaces are not level, but often much rounded for drainage, and truck suspensions are not rigid. The van, like the plane table, must be leveled.

NEXT MORNING DOWN THE ROAD WE rolled. It was commencement day for the local high schools in these intimate little towns, and every local radio station we heard warmly congratulated each new graduate by name, one after another, all morning long. The road was really straight; the only curve we had taken before lunchtime was along the east bank of the Calamus River. There the highway first bent leftward and after a dozen miles or so bent back to the right by the same amount, to bridge the Calamus and straighten out southward again. That would be an easy correction.

We crossed the wide River Platte, and soon after that entered Kansas across Prairie Dog Creek in mid-afternoon. There was only one more offset like the Calamus all day—plus one experience we had forgotten about, one always lying in wait for motorists: an unmapped detour we reached toward sunset. A bridge was out. The temporary detour was well-marked by the Kansas highway people, and we could easily adjust our mileage log for it. We needed to

allow for all the extra miles we drove that were not in the direct line south. Hills were few and low; it was easy to figure out that they could be ignored in finding the distance. We had checked the van odometer carefully against the road map for some miles as we were driving to Bassett. We were satisfied with the distance result; the biggest error would be, we thought, in the two sightings on the star.

WE SET UP AGAIN AT THE AIRPORT IN COLD-water, Kansas, a long day's drive south on 183 from Bassett. We made sure the van was level and aligned with the highway, and waited again for Antares at the dark airport. This night was pleasantly warmer, and above all the sky was wonderfully clear; it was a joy to watch the stars, and Antares shone out as it came across the sight on the van. The most important precaution for this sort of astronomy had become evident from experience: pick a really clear night! The second tape was placed on the wall of the van, and we returned tired and content back to the motel a little before the dawn. The calculations had to wait, though we were eager for the result.

We drove out to the park later in the day to work it all out under the blue sky. The two lines of tape on the van made very plain that Antares had arced higher in Coldwater than it had in Bassett. But the star is so far off that it must in fact still lie in the same direction. What had

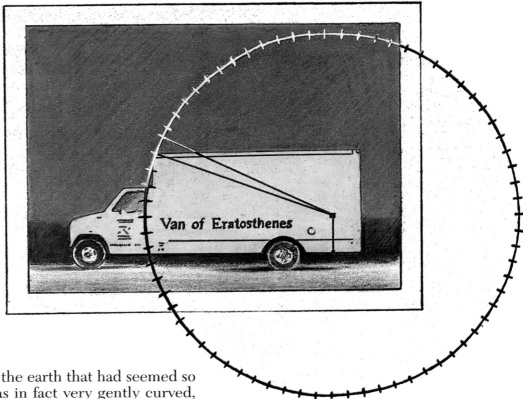

The small arc on the van is part of an entire circle on the diagram. The small arc we have driven on the flat-looking plains is the same part of the entire circle around the earth. Just under seventy-two repeats of our journey would take us all the way around.

happened is that the earth that had seemed so flat in Kansas was in fact very gently curved, and we had driven down around that long slow curve all day. If Kansas were as flat over hundreds of miles as it looks to be locally when you can see only a few miles, the distant star would have been in the same direction from the van at start and finish, and the two marks would have merged to one. It is the roundness of the earth that is marked right there on the van wall, for the plumb line, which formed the other side of both the sightlines, pointed to the center of earth at each sighting.

We had planned how to transfer the angle marked on the van to a worksheet. We placed an arc drawn on paper at some distance from the pegs, and marked the spacing between the tapes on that arc. Then we laid out on a picnic table a big sheet of paper with a whole circle drawn on it to the same radius. Now it was easy to transfer the tape spacing to dividers, and to walk out the dividers around the whole circle counting step by step. That arc was in proportion to the long day's drive as the whole circle was proportioned to the entire distance around the earth.

The arc could be laid out over and over again just seventy-one and three-quarters times to round the entire circle. We had pored over notebooks and maps at breakfast to adjust the van's odometer reading for the few curves and the detour, and we had found that the distance traveled straight south from Bassett was 370 miles. Plainly the whole earth is then 71.8 times 370 miles around. A small calculator did the multiplication: by direct measurement, courtesy of trusty van and of far-off uncaring Antares, we make it 26,500 miles around the earth.

We had rounded off the number, of course, for ours is clearly no precise way to sight a star, or to measure a road in miles. But our amateur expedition was audacious enough—and great fun—and the best methods of our day give about 24,900 miles for the earth circle, about six percent smaller than our result. Not bad!

I wonder if this is not a rewarding problem for other amateurs of science, young or old? A van on Highway 183 worked well, but it is not indispensable. The earth is just as round in Maine or Arizona or even on the surface of Lake Michigan as it is on Kansas plains. What is indispensable is to understand and to improvise a method, and then maybe talk friends or family into one overnight trip.

WE HAD WATCHED THE ARC FOLLOWED BY the star Antares change noticeably between Bassett and Coldwater, as we drove south straight toward the high point of its arc. It is easy to conclude that the star path would differ on other north-south trips only by shifts of just the same sort, simply making a higher arc if we had kept on southward beyond Kansas, or a lower one if we had turned the other way, to head north from Nebraska.

Start anywhere—Tokyo, New York, or Delhi, let alone Bassett, Nebraska—and the same sorts of changes modify whatever star paths you see. The stars will arc higher or lower as you move, always by the same fraction of a circle for the same distance in miles. That uniform result makes clear the roundness of the earth in this section along a north-south line.

When you move due north or due south you can always know how far, by checking any stars mapped in the sky as they arc past the north-south line. But if you move east or west instead, every star's path remains unchanged. Each star

follows the same arc; it is only the time that it reaches the top of the arc that changes. Go east to meet the rising star, and the star climbs to the top earlier; west, later. So a change of position east or west cannot be found from star paths alone; you need to know something about time.

Latitude is a term for the distance of any point from the equator, measured not in miles, but as a fraction of the circumference of the earth. Places which lie at the same distance from the equator are on the same circle of latitude. (It is an old convention to measure latitude in degrees—the circle has been divided into 360 equal degrees ever since Babylonian days—and to count them either north or south from the equator of the earth, which is midway from pole to pole. That zero for latitude was chosen with a sense of symmetry; either pole would work as well.)

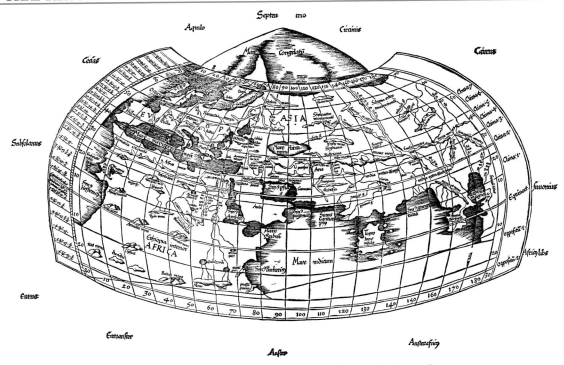

The Greek world as mapped by Ptolemy of Alexandria. All that Renaissance Europe knew of what the second-century geographer had done was his tabulation of many positions by latitude and longitude. People compiled maps from those positions to illustrate the old book; this map is from an edition of Ptolemy's *Geography* done in 1522 in Germany.

ERATOSTHENES

ERATOSTHENES, WHO HAD BEEN educated in Athens, was chief librarian of the Museum of Alexandria, its learned director of research, somewhat before 200 B.C. His claim to earth measurement is well-founded, though the result he got is less clear because we are not very sure of the size of the units in which he expressed the result. His own interest was rather in method than in result; he may never have seriously tried the scheme.

He used the sun, not the stars, but otherwise his method was close to what we did with the van. The Nile runs pretty nearly north-south; his city was near the mouth of the river, and far to the south was the town of Syene. The sun is not so regular as the stars. But on one day of the year, midsummer's day,

when the sun was highest, the sun passed right overhead in Syene. Everyone knew that; it was marked by the reflection of the sun off water in the bottom of a deep well.

It was easy to measure the shadow of a small vertical rod when the sun was at its highest on midsummer's day in Alexandria. The diagram he could draw was like the one we drew for the two positions of the van; the sun, like the stars, is far away. The shadow of his rod gave him an estimate for the angle by which the sun missed being overhead; he judged it the one-fiftieth part of a circle.

The road distance south to Syene (called Aswan today) that Eratosthenes used was almost certainly paced off by professionals for the Egyptian state who walked between all important places. There is historical evidence for such a craft; road distances are useful to planners with military responsibilities for any good-sized kingdom. The value Eratosthenes cites was 250,000 stades around the earth. Scholars cannot be sure of what a stade meant then, but the result is not bad on any likely guess for the length of the stade.

THE ASTRONOMICAL TECHNIQUES OF MEAsuring elevations above the horizon were not Eratosthenes' own; they were older even than Greek science. After Eratosthenes the problem attracted other scholars and their patrons all around the world. The root idea is the same as with our rental van: if you measure the elevation angle of sun or star from two different positions, you can relate the circle of the earth to the physical distance you have stepped off north and south.

Here is a list of some early players of this game:

The Chinese measured a north-south line at least a few degrees long on the great plains of the Yellow River about 720 A.D., under the T'ang dynasty. They especially wanted the shadow angles of the midsummer sun.

About 820 A.D. the caliph al-Ma'mun caused arc and distance measurements to be made around Baghdad and Damascus by the scholars of his academy, called the House of Wisdom.

The young Jean Fernel, who became a celebrated physician, measured a degree by his own personal efforts between Paris and Amiens in about 1530, counting the turns of his carriage wheel to fix the distance. His was the first try in Europe.

Willebrord Snel (of "Snell's Law"), sighting out around Leiden big flat cross-country triangles that connected his own house with many a church spire, found a good result from one baseline in the year 1615.

Richard Norwood stretched a chain of links again and again all the way from London to York and measured the change in sun angle, about 1635.

Finally, the group of astronomers who were later to be brought into the Paris Observatory measured accurate arcs about 1670, both abroad and in France. Jean Picard introduced the telescope into surveying instruments for the first time, a powerful improvement.

The Paris geographers became much interested in small differences that might indicate an off-spherical shape of the earth. A small flattening of the polar regions of the spinning earth was predicted by Newton on theoretical grounds—"the Planets are something thicker about the equator than about the poles"—and indeed the disk of Jupiter showed such flattening in the telescope. Direct measurement over the ground is difficult, and was not carried out conclusively until the middle of the eighteenth century, after French expeditions to Peru and to Lapland.

The local view.

"Not in Kansas Anymore, Toto"

EVEN TOTO MIGHT HAVE GATHERED THAT already from the wide stretch of flowing river that ended every open view. We will not tell you where this is; we came here to find that out, placing ourselves correctly on the map of the earth by using the starry globe of the sky.

This part of the chronicle carries us to the end of the eighteenth century. The world map was as popular then as satellite images are today. Some of that sense comes to us from the handsome little pocket globes of the earth made in that time, prized by collectors today, and sure to be in every museum. My own three-inch globe is a deft piece of work signed by the craftsman Nicholas Lane in London, about 1780. It is a good map of the whole earth as they knew it then. (One map entry near Hawaii bears the news of the death of Captain Cook, the most celebrated of Pacific explorers.) Every young person in a good family had then to study such a globe, to learn the parts of the earth. We still do that. Our libraries and our schools quite traditionally have globes of the earth. But in those days they did something more with a globe, something quite revealing that we have almost forgotten. You expect a pocket globe to come with a neatly fitted case. It is here, but the case of this globe is not merely a fitted black container. The inside surface of this spherical case is another globe, the hollow sphere of the starry sky, into which the real earth fits as intimately as the pocket earth fits its own small blue star-sphere. They still celebrated that relationship in those days; we tend to forget it now, even though our map making remains largely based upon it.

WE CAME TO A PLEASANT GREEN RISE IN THIS town. They call it Ellicott's Hill. Andrew Ellicott arrived here in February 1797 to carry out a serious piece of mapping. Ellicott was then the most expert surveyor in our young country; a few years earlier it had been Ellicott and his partners who laid out on the ground the newly planned District of Columbia. Ellicott had traveled half a year to reach this place, then only a frontier village a long way from his Pennsylvania home. He held an urgent commission from President George Washington. New owners of real estate are often a little anxious about their boundaries, and the new United States was no exception. Ellicott was sent to travel our national boundaries to make sure they were well-mapped, especially here at an uncertain border with another nation. He spent a year here or camped nearby, on a mission that linked touchy diplomacy with skilled mapmaking.

Once again we intended to carry out the work of some old mapmaker by methods resembling the old methods, not out of a merely antiquarian interest, but because their real solutions to their real problems are often simpler, understandable antecedents of our

Our three instruments, clock, transit, and telescope.

A brass pendulum clock, rather like Ellicott's, but made recently.

An aging transit of the twentieth century.

own elaborated methods. So we made our way to Ellicott's Hill, carrying mapper's instruments, none of them true antiques, yet all close relatives to the ones Andrew Ellicott brought here once to put this place on the map.

Three instruments came along with us by plane and car. First was a brass weight-driven pendulum clock, handmade a few years ago by Boston clockmaker Jim Moss, who carried it here for us. Ellicott had carried just such a clock, one he had built himself, made to a pattern quite similar to Moss's modern example. Then we brought along a serviceable transit, a generation or two old, a low-power telescope mounted on graduated circles to measure sightlines. Ellicott owned a beautiful transit he had constructed for himself and had put to use, as he says, in laying out the "principal avenues in the city of Washington." He and his well-equipped party had brought as well a number of telescopes without any circles, not intended for angle measurements, but choice for star-watching. We brought one, too.

Let us place Mr. Ellicott's hill on the map more or less as he did it. We had to manage in one evening; he spent months at the task, repeating and checking, measuring both by day and by night. First, we measured once more the high point of the arc of a bright star, just as we did in Nebraska and Kansas, but more accurately than sighting along the wall of our van.

The small transit telescope let us measure fainter stars, so we had a much wider choice than bright Antares and its few peers. We picked another star that is well located in the sky and found its maximum arc height, using the cross hairs of the transit and reading from the finely graduated circles the angle to the star. Once again our tool had to be leveled; it had several bubble levels rather than a plumb

Ellicott's instruments.

line. The astronomer's tables then gave us our latitude, once we knew the mapped position of the star and the measured height of its arc.

Our value for this latitude: 31 degrees 32 minutes north.
Mr. Ellicott's 1798 value: 31 degrees 34 minutes north.

We found our north-south position, as he did, quite well, within a mile or two on the round earth. The conventional degree is one part in 360 of the entire circle of the earth. A difference of one minute in latitude (there are sixty minutes to one degree) therefore means an uncertainty on the ground of a little above one mile. Ellicott measured three stars repeatedly, using his transit and another big and more

specialized instrument. His old work is better than our quick effort, closer to what the modern maps of the U.S. Geological Survey give for the place. We do not know for certain where Ellicott observed, but it was on this hill, certainly within a mile or so, if not right on our very spot.

This measurement alone places us on a particular circle on the earth, located at 31 1/2 degrees north of the equator. (We almost share that circle with Alexandria and Shanghai, both distant great cities lying about a dozen miles south of our circle.) Yet we do not at all know where along that earth-girdling circle we are. Are we near Shanghai? The stars alone cannot tell us that. They will look just the same this night in Shanghai as they do here; so will the sun by day. We need a suitable clock if we are

STANDARD TIME, SUN TIME, STAR TIME

Time measurement is an essential for east-to-west position on the world map. We all habitually measure time in a conventional way. Your wristwatch keeps "time-zone" time, Eastern Standard Time, Pacific Daylight Time, or whatever it may be. That is a convenient scheme for a fast-traveling world, but time by zones is not directly usable for mapping. Eastern Standard Time is kept to the split second by people all the way from Michigan to Maine, though it is really an averaged sun-related time best suited to Utica, New York.

The time zone—if you know it—does offer the germ of the idea of east-west positioning on earth. No Easterner who has watched sports on TV can have missed the sight of the game going on under the bright sun in California while the eastern states are under darkness. The Pacific Coast is three hours earlier by sun and by stars then is the Atlantic. That is an eighth of the way around the world, one-eighth of the full twenty-four hours it takes the sky to pirouette once on its axis.

Sundial time, on the other hand, is now-adays an anachronism, but it is genuinely local time at the place where the dial stands, quite workable for the mapper. But it is not simple to use unless you are nimble at arithmetic. The sun has a complicated motion in our sky, appearing high and low, fast and slow over the seasons. We prefer to stay away from the series of calculations or tables that the sun's awry motion requires.

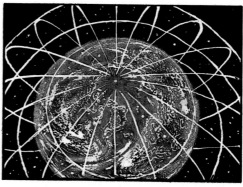

No, we propose to map by star time, the simplest and most ordered choice of all. That is the astronomer's time, a night-oriented scheme. The stars all come by as we

The star dial diagramed, first as it looks from the ground, with imagined lines marking the stars as the sky turns. Then we move it in thought out above the pole of the earth, so that indeed it looks like a flat clock dial, the dial for star time (or sidereal time, to speak more formally).

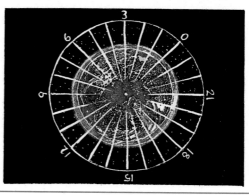

watch the smoothly turning sky: that starry sky can be thought of as a dial, which turns exactly once in twenty-four hours of star time. (Its uniform round defines the day and the hour of star time.) If any star arcs high, crossing the south as you watch, that same star will stand there again just twenty-four hours of star time later, season by season, year by year. The sun is not at all so obliging; star time and sun time diverge and rejoin each other over the four seasons.

To use the stars as a clock dial, there must be conventional marks, like the numbers on a clock face. Simply divide the sky into twenty-four equal sectors, and each will span one hour. All we need do then is assign to any star some particular time it can carry across the north-south line, as for sun time the sun carries 12:00 noon. But there are many stars, and none dazzles like the sun.

Long ago a zero hour was chosen on the star dial, at a position that is really arbitrary in this context, although it has a clear astronomical meaning. Once that zero was picked, any map of the stars assigns to every star its own time, which follows simply from where the star is on the dial—that is, in which hour and at what fraction past the hour. For example, the entire figure of Orion lies between 05:00 and 06:00 on the conventional dial, and bright blue Rigel, his left foot, is at about 05:12. Whenever Orion arcs due south of an observer, he can read off his local star time as between 05:00 and 06:00, and Rigel provides a minute hand. That is the very same logic (though not the same time) by which we assign 12:00 hours, local noon by sundial time, to the moment when the sun stands highest for the day.

to find our position east and west under the turning sky, and so did Ellicott.

We used the sky itself. As you watch the night sky, the hour getting steadily later, stars that bear later and later star times cross the south direction. Every star there will troop past dutifully each twenty-four hours—of course, daylight glare hides the ones that are too close to the shifting position of the sun.

The star we used is called Spica, a bright and conveniently placed star in the constellation Virgo. Its time mark is 13:24 hours, since it is somewhat past the halfway mark on the star dial. If anywhere in the world you catch Spica at the top of its arc, your local star time is 13:24.

That is how we set our own star clock. It does take some care to judge when a star is at the top of its arc: there its path is nearly horizontal, so one must watch a little beforehand while it is coming up, and a little afterwards going down. Ellicott used the symmetry of the star arc in a clever old trick to get still better timing, and we followed his lead. It took some time to find the result; with our pendulum clock we allowed for the delay.

But this is still not enough. To find where we were on the circle of latitude we needed one more piece of information: what we wanted to know was how far we were east or west from any well-established place, say Washington or Greenwich or Paris. With that information, we would be located on the globe.

We could learn that simply by learning what the star time is somewhere else, for then we could calculate how many hours and minutes of the turning sky lay between us and that place. If at some spot the star time were six hours earlier than our star time, then we would know that we were west of that spot by six hours out of twenty-four, and thus by one-fourth of our full circle of latitude around the earth. That would put us west of the reference by 90 degrees of longitude, as east-west difference is called.

I dialed direct from the old tavern here on Ellicott's Hill, built only a few years after the commissioner had passed by, to my astronomical friend in Paris, Suzanne Débarbat. Mlle. Débarbat is senior astronomer and historian at the Observatoire de Paris. I knew she had an observatory star clock close at hand, and she

The time comes in from Paris. The Paris star clock and our own gave us how far the stardial had turned between Paris and the place where we watched. We were in Natchez, Mississippi.

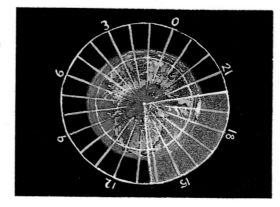

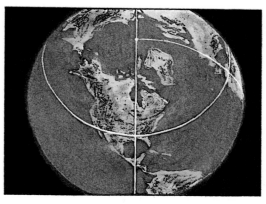

had generously agreed to help us. The full flavor of our brief conversation comes out on listening to a selection from her side only.

DÉBARBAT: *Allô?*

Yes, certainly…Yes, *bien*…5, 4, 3, 2, 1…top! At the top it was exactly 20 hours, 2 minutes, Paris star time.

Now we could put ourselves squarely on the map. Paris read 20 hours and 2 minutes "at the top," when our star clock read 13 hours 47 minutes. The delay since Spica crossed just south had been kept track of by our pendulum clock.

So we had to be west of Paris by the amount the sky turns in 6 hours and 15 minutes, a little more than 90 degrees away. We wanted to compare directly with Andrew Ellicott's result. But his reference was Greenwich. We dug out of the *Nautical Almanac*, where there is a long table of the positions of all the observatories, the fact that the Paris Observatory is situated east beyond Greenwich Observatory by 9 minutes and 21 seconds of time.

A bit of careful addition and subtraction, and we had found our position, placed Ellicott's Hill for ourselves on the map of all the world. It lies on the latitude circle we measured earlier, and west of Greenwich by the distance just worked out. The two measurements put us on the map. On the map this is the global address

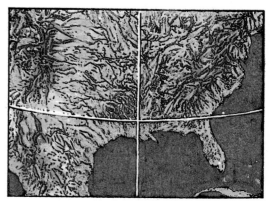

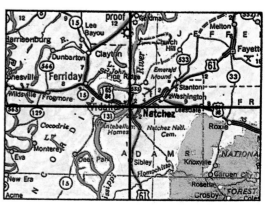

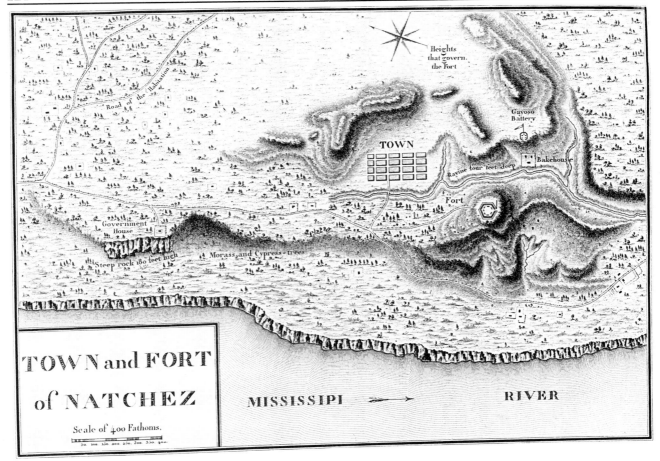

Frontier Natchez in 1803.

of the attractive riverbank town of Natchez, Mississippi. The river itself flows broad and deep only a quarter of a mile away, well below our hill.

Our results in 1986:

latitude: 31 degrees 32 minutes north

longitude: 91 degrees 25.5 minutes west of Greenwich

Commissioner Ellicott found in 1798:

latitude: 31 degrees 34 minutes north

longitude: 91 degrees 29 minutes west of Greenwich

The results are pretty much alike. Real measurements seldom repeat themselves exactly;

you always hope your answer is a little more reliable than it is. At Natchez each minute of latitude or longitude works out to about a mile on the ground. Not bad! We will return to this hill once more, to locate it on the map by the most modern and precise of all mapping methods.

ANDREW ELLICOTT CAME TO NATCHEZ LONG ago to begin formal demarcation of the boundary between the new United States and the domains of Spain. Natchez was on that boundary. That was why Mr. Ellicott did such a thorough job of locating what was then a tiny fortified village.

But how could he hope to do it? When we did it the essential step was to gain timely information from a distant clock. But Ellicott had no telegraph, no telephone, no radio, not even

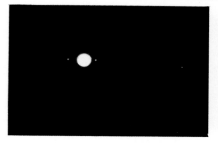 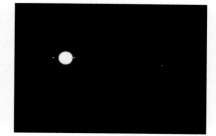 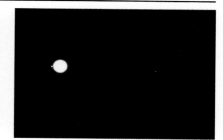

The disk of Jupiter and the three closer starlike moons circling the planet. The photos taken all in one night show something of the motions, and one eclipse, when a moon ducked into the shadow of the big planet.

the indispensable credit card! There was then no means to send a time signal of any sort promptly enough to give him the local star time at any well-mapped place.

One approach that became of wide use during the generation or two after his time was simply to carry an accurate clock portable enough to keep running on shipboard or stagecoach. You could set it by Greenwich time whenever you left any place already reliably on the map with respect to Greenwich. That was not possible with pendulum clocks. They won't keep time at all in transit, whether by rocking stagecoach or rolling ship. Jim Moss had brought his pendulum clock lying asleep in its case, not running at all. For mapping with good results you certainly need timekeeping good to a minute or better. In those days you had to plan on perhaps weeks or months of travel. Pendulums will not do, and most portable spring-wound clocks or watches cannot keep good enough time. There had just been developed in Britain and France elegant small spring clocks of high accuracy—they are called chronometers—but they were simply too new and too dear in President Washington's America.

There was one way. Galileo had shown nearly two centuries earlier that if you looked through a modest telescope at the right place in the sky, from almost any point on earth, from almost any day in the year, you could see and read the dial of a single great natural clock in the sky. That was why Ellicott brought a telescope here, and that remote clock in space was the clock that Ellicott used.

The Map of the World

IT WAS IN THE SPLENDID NEW OBSERVATOIRE de Paris, built under Louis XIV in the year 1671, that the astronomers first calibrated and made useful the global clock in the sky. The idea had been Galileo's, but he was never able to reduce his proposal to practical form. Long after his death, the brilliant staff of this new institution succeeded in that task.

That clock in the sky is the planet Jupiter. Around Jupiter dance four bright Galilean moons, easily seen in a modest telescope, an instrument not hard to carry on your travels. (The scheme fails for shipboard; a ship at sea is too unsteady a platform for a telescope of the needed power. The events that can be seen among the moons are rather infrequent; you often must wait a few days for one. That is unsuited to the navigation of a moving ship.)

Jupiter is so far away from earth that its moons appear the same whatever your location on earth. It is necessary to work out the rhythms of the motion of the moons well enough so that their future positions can be predicted for a reasonable time ahead, say a year or more. What people watch for in the tel-

Louis XIV, the Sun King, paid a real visit of state to his new academicians of science in their Paris headquarters in 1671. The engraving records the very event. The four visitors in royal dress include the King. Behind them are the less spectacular academicians, all recognizably drawn among their books, specimens, and apparatus. Through the window can be seen the new Observatory under construction; the artist has taken the license of moving it a mile or two north, to bring it into his delightful picture. (J.-D. Cassini is the man seen speaking over his shoulder, just below the skeleton ribs of the stag.)

escope are times when any one of the moons passes behind the round disk of the planet to be hidden, or times of eclipses that cause the moon quickly to vanish in the big planet's shadow. Such events can be timed to a minute and better. First the astronomers at Paris had learned to project and publish coming moon events well ahead. Then anyone who had their up-to-date tables could read Paris time from the circling moons of Jupiter on most nights of the year. The clock is the Jupiter system; its hands are the moons; its dial is the published forecast of their motions. But you can read it only with a telescope.

Finding your place from the clock in the sky works this way. Watch by telescope for any predicted event among the moons of Jupiter, wherever you are. Then read the time in Paris directly from the Paris schedule of moon events. Compare it to your local star time (or sundial time, as you prefer). It is not easy; one minute of uncertainty in time implies about ten miles uncertainty in location. But it worked well in expert hands, and in two or three decades it enabled the compilation of the first modern map of the world.

The documents and the instruments that tell the story of that map are kept in the Observatoire de Paris, behind its fine classical façade ornamented with carvings of seventeenth-century instruments of mapping and astronomy. (Inside the building I was pleased to find more than one clock handy on the wall whose dial promised *temps sidéral*, French for star

Februarius. 1668.

Configurationes Medicæorum.

Hora 7. P.M.

How to move to Paris. Cassini's 1668 tables of motions of the moons of Jupiter were the best. He could predict the events a few months ahead, using a mathematical account of the rhythmic dance of the moons. We have drawn a sample curve right over his published diagrams.

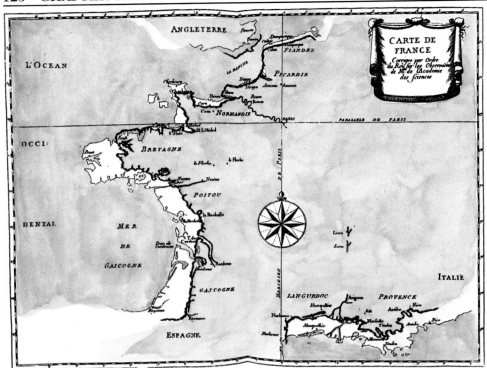

France was as well mapped as any other country before the astronomers tried their hand. But the old travelers' methods were simply too rough. The coast of France lies only a couple of hundred miles west of Paris, yet when Cassini and the rest first pinned the map of France to star time using the moons of Jupiter, they found that the older maps had put that distance too far by about sixty miles. France was in fact smaller than the map of 1679.

time.) Suzanne Débarbat and the librarians there generously guided me in examining the treasures in all detail.

The name most closely associated with the map project is that of the astronomer Jean-Dominique Cassini (whose son, grandson, and great-grandson would all in time become luminaries of the Observatoire and of French cartography). I was invited to pore over the pages of a famous book published in Bologna in 1668, written in Latin by a brilliant man of forty-three, who was then also engineer to the Pope. His name was Giovanni Domenico Cassini; of course, it was the same man. That book was a set of his observations and tables of the moons of Jupiter, and it ended with a couple of months' precise forecast of coming events among the moons.

In Paris the astronomers of the Académie Royale des Sciences were already at work, if not yet at home in the new observatory then under construction. The tables from Cassini were exciting in their quality, and a careful check of his forecasts convinced the academicians that here was the man to attract to Paris for the big discoveries and the grand mapping projects of their hopes.

Before long Jean-Dominique Cassini had become a new French citizen, who lived with his family in the fine apartments of the Observatory. His colleagues were no less illustrious; they included the great Dutch physicist-astronomer Christiaan Huygens, and the Dane Ole Roemer. There were a number of accomplished French scientists like Jean Picard, who had just completed surveys that for the first time used the surveying instruments he had equipped with small telescopes and carefully graduated circles. His innovations improved the accuracy of sightlines by tens of times.

By 1679 the Paris astronomers were ready. They began to remap France by the new methods, setting up in town by town along the coasts, Dunkerque, Calais, Cherbourg, Brest, Bordeaux, and more. Their new map astonished the king, who could see that on it the Atlantic coast of his realm had shrunk fifty or sixty miles closer to Paris than the best old maps had put it, that much change out of only two hundred miles or so. Mlle. Débarbat quoted the remark Louis made then: "I paid my academicians well, and they have diminished my kingdom!" It must have seemed a little unfair.

I think the king was only joking, for the new maps whose framework was given by Picard's transits and Cassini's tables of the eclipses of Jupiter's moons made France easily the best-mapped land in the world. The locations on the map begun in 1679 stand up to present values within a couple of miles. About the same time Cassini and his team began ambitious plans to map the whole world in the same way, training eager, mathematically minded young Jesuits so that they could observe once they had reached their missions abroad, writing manuals of instruction and finding instruments for them, even mounting an expedition or two from the Observatory itself to test and support the scheme.

It was on the floor of the beautiful eight-sided room in the west tower of the Paris Observatory that Cassini once plotted his map of the world in circular form, what he called his planisphere. He entered the North Pole right in the center of the floor, and placed all the other points by laying off their angular distances from the pole and east or west from the north-south line through Paris. A fine parquet floor is in that room now; underneath the wood on the smooth limestone blocks the archeologists might some day make out the faint inky traces of Cassini's planisphere.

The cities and landmarks from which the data flowed in were entered as the news arrived, the best points measured by star transits and telescopic timing of the moons of Jupiter, the very method still used by Ellicott in Natchez a century later. Astronomers all over Europe sent in their own well-checked positions. The Paris-trained Jesuit missionaries reported from far-flung posts; Quebec City, Santiago de Chile, Guadeloupe in the West Indies, Lisbon, Venice, Cairo, Cape Comorin at the tip of India, Siam, Canton, Peking were all there. They were marked right on the stone floor as they came in, to form a master map that grew over the years.

That map on the floor was transcribed and published in 1696, the first modern map of the world, the first to be based on a framework of reliable positions. The certified entries were marked with an asterisk. About forty such places had been located with control from the Observatory. Those locations all over the world differ on the average less than fifty miles from what we would plot today.

Coastlines and landmarks far from centers of population were not yet so firm. No one had mapped them by the methods of Paris. Their mapping depended still on the rougher estimates of travelers and shipborne explorers, who had no way to measure longitudes at all. Such reports were not yet tied to the sparse points of the framework. Notice that the sailors had done some guessing; California, for example, is imaginatively drawn as an island. Some regions were hardly known in Paris at all, like the northernmost Pacific, nor anything very far south toward the icy continent of Antarctica.

But the way to the map of the world was opened wide.

THE FIRST TRUE MAP OF THE WORLD

IN 1696 CASSINI HAD THE PLANISPHERE on the floor of the Paris Observatory made into a printed map of the world. This was the first world map to have a firm framework pinned to the sky. Reliable positions were known for about forty places all around the world, which we have entered here in red. Latitudes were referred to the equator, longitudes not to Paris, but to the island in the Canaries which was traditionally the westernmost point mapped by the Greeks. Coasts and details were still uncertain at some distant places; look at Australia and the Pacific Arctic. The forms of the coastlines appear unusual, partly because they were not always well known, but partly because this circular projection is not very familiar.

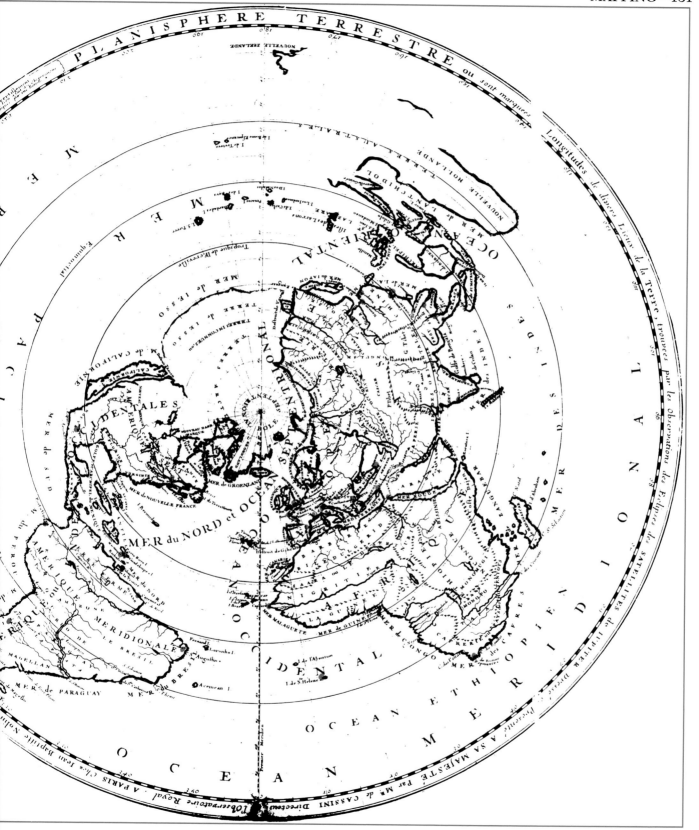

Even handsome topographic maps are no better than their control points. Aerial photography and surveys on the ground have supplied the detail recorded here on a colorful, complete, and helpful map, but the control points are still its frame. The benchmark shown in the previous figure, benchmark No. 10, is entered near the center of this small portion of the U.S.G.S. map named San Francisco North.

Pinning down the map, this official bronze marker is one among close to a million that fasten the map to the world itself. In between control points, map details can slip and slide. This one helps fix the shoreline of San Francisco Bay; its location is well-measured, like the forty places starred on Cassini's Paris planisphere.

THIS BRONZE MARKER PINS THE MAP OF THE world at one point to the earth itself. When they made the first modern map of the world in the Paris Observatory long ago, they could pin their world map down in only a few dozen places, where a trained observer had carefully measured the address of some city or landmark around the world. Nowadays we can count those points by the hundreds of thousands, in all well-mapped countries. This one belongs to a set of a few dozen that together outline only the shoreline of San Francisco Bay. Whenever you want to make a good map that is reliable in scale and form, one that can be fitted into the map of the world, you need to begin with some well-known control points. For in between such pins, your map can slide and stretch or even tear.

These are the bones of the map, its hard framework. A map is a good deal richer than a scattered set of pinpoints, but those pinpoints—the control points—are what give it firm shape and size. Most of the colorful markings that make up a highway road map or the map in an atlas are drafted by hand, or nowadays by computer. That is how the curving lines and crowded symbols and varied lettering actually come onto the printing plate that laid your map down on paper. But the points are certainly there, sometimes indicated, sometimes only behind the scenes.

Form and detail for a map these days usually begin with photography from the air, repeated over the years to allow maps to be updated; whenever man or nature changes what is on the ground, the map must follow. Overview by airplane is far quicker than living up to the honorable tradition of walking the ground with plane table and sketchpad.

A photo from the air is not yet a map. It is only the basis for one. Even if the aerial camera lens were quite free of distortion—it is pretty good, but not perfect—the airplane drifts, tilts, and veers as it flies along, and its height above-ground can hardly be kept entirely constant in flight. The ground below is not to be squarely caught in any airborne camera. It is as though we tried to fit our sightlines on differently tilted plane tables. The slightly awry camera images then require correction, for we count on a map to give a correctly formed view that is

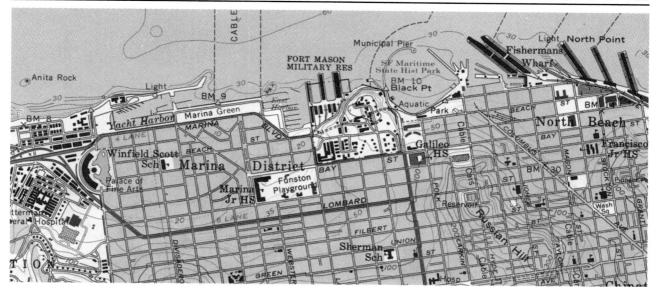

everywhere to scale as seen from above, and not the attractive but out-of-scale perspective landscape of the artist or the landscape photographer.

The quantitative handling of images is a swiftly changing and advanced technology in this digital age. One widely used and understandable procedure—though not the only one—begins by effectively carrying out the aerial photography in reverse: the photo taken from the air is projected, at a variable size, onto a flat screen that the technician is able to adjust in position and slope, until the projected image is best in form and scale. That judgment is made by fitting the projected image at the right scale to a number of recognizable features related in position to the ground control points. At the marker we first saw, for example, the mapper would fit the aerial view to match the road corner where he knew the small bronze disk was set. The map is built up as such correctly projected outlines are traced for a whole mosaic of air views, photo after photo. All the crowded and valuable symbols are then added, the political boundaries, the areas marked in color after color to show the use of the land, the nature of the roads, the identity of the public buildings…

But the most colorful and complete of maps will never be more reliable than the marked and measured points that make up its control framework.

Mapping by Time

UP TILL NOW WE HAVE REGARDED MAPPING the world as mainly a visual matter. It is wholly based on sightlines to land or sky, all of them some version of looking out to just where you are pointing. Taking distance by stretched cord or rolling wheel (like our van across the prairie), finding directions by direct lines of vision, checking vertical and horizontal by a hanging weight or by a bubble rising through a liquid—they are the fundamentals.

But more and more the world is being mapped by a method that you can imagine doing with your eyes closed. Suppose we knew of a runner very steady in performance who ran moreover always in a straight line. By timing his run we could know how far he had come. That turns out to be enough so that we can eventually reduce the mapping of the world to the measurements of time.

One such runner—not the only one—is sound in air. One delightful morning of blue sky and scudding clouds I stationed myself at one goal of a soccer field. Over there at the other goal across the soft green grass was a source of pulses of sound: a rhythmic and untiring trap drummer. Each single drum tap sounded clearly from the far goal, one after another.

She and I each had a two-way hand-held radio, what people call a walkie-talkie. With

A small spinning motor easily splits the subseconds it takes sound to echo back up. The watery drawings show the sound leaving the boat at the surface to rebound later from the bottom. The other column shows the way a contact on the spinning shaft keeps track of the time delay, flashing a light wherever around the dial it is when the amplified echo returns. A depth of twenty-five feet means an echo delay of a hundredth of a second.

the radios on, I heard every single drumbeat twice. Its sound came first over the walkie-talkie, coming by radio without appreciable delay. Then very quickly the single initial beat sounded for me a second time, this time coming directly through the air at the speed of sound. There were always two distinct beats, *tap-tap*, well-separated to any ear. Once you learn how to measure that time delay you can associate a distance with the delay, and use time to map the world.

Nimbler rhythms than drumbeats in the ear are beginning to replace the sightlines of the keenest mapping eye and telescope. For nowadays the timing abilities of the ear are wonderfully extended by split-second electronics.

On Vineyard Sound off the mainland of Massachusetts even the sport fishermen who go out for a day trip in a small boat are equipped to come back to the right spot, by timing echoes in order to feel out the form and depth of the bottom. Sound in seawater travels a little less than a mile in one second, much faster than it does in air, where its speed is about a thousand feet in a second, or twelve miles a minute. No one has found a cooperative mermaid on the bottom to tap a drum; if you want to pick up any sound from the bottom, you must generate it there yourself.

The natural way to receive sound coming up from the sea bottom is to start an echo from there; the instrument on the boat makes a strong sound pulse (usually at an inaudible ultrasound pitch) from an underwater loudspeaker aimed straight down. The sound travels down, reflects at the bottom (sometimes even from a school of fish partway), and a weak echo eventually returns to the surface. A microphone on the boat picks up the echo on arrival to be amplified and timed. In these coastal waters the relevant round-trip time is a hundredth of a second for each twenty-five feet of depth. An inexpensive device based on a spinning motor is quite able to split time that well; the little motor, not much different from a small electric fan motor, turns a contact disk at the rate of fifty or sixty turns per second. It is not hard to arrange the electronics so that a small light that turns with the motor shaft flashes whenever the echo comes back. The depth is signaled by the dial position of the repeated flash as the boat sails on.

With more powerful sound gear and better amplifiers, you can arrange for the return echo to move a pen up and down a recording chart of paper that unrolls as time goes by. If the pen motion is scaled to the amount of echo delay, it traces a profile of the bottom as you pass. The position of the sea surface is steadily marked by the sound of your own engines, always present as a noise in the detector.

We went along on a larger boat with a good echo profiler to look for an old wreck on the

Satellite mapping at Ellicott's Hill. The computer and superb clock are in the small boxes; the antenna and receiver are a little plastic hatbox on a tripod a dozen feet away; the precisely ticking satellites are ten thousand miles out there in the night sky.

bottom of the Sound. First we found the nearby dunelike hills of bottom sand that the tidal currents past the wreck had built over the years, and then careful searching picked up the abrupt rise as we crossed right over the hulk of the old freighter, invisible on the bottom below.

The oceans have been measured to all depths—after all, the deepest ocean trench is six miles, with echo time a dozen seconds—by powerful profile recorders kept at work on oceangoing ships ever since the latter years of World War II. A million reliable depths are known and plotted worldwide; of course, if a depth is to be put on the map there must be a way to measure the ship's position as it sails. That, too, is usually done by timing, not with sound in air or water, but by one or another radio technique. Radio can span great distances at its speed, which is the ultimate speed, a million times that of sound in air, or a thousand feet in one microsecond. Thus, radio (or light) travels a thousand miles in the same time that sound in air advances five feet.

To locate a position on earth to a matter of a mile or so, timing must be good to microseconds. That has been pretty easy for the last fifty years because of electronics. The map of the ocean depths is the product of such timing.

The famous expedition of the ship *Challenger* in the 1870s was the first to carry out many soundings of the ocean depths worldwide. In those days the ship's position had to be found, not by radio, but from sights on sun and stars, using a batch of accurate shipboard clocks, the chronometers. The depth of water was sounded by a heavy weight that sank slowly on a few miles of one-inch hemp rope. A depth measurement was a considerable part of a hard day's work at sea. Usually samples of the water and the bottom were fished up along with the line. In her thousand days around the world HMS *Challenger* measured ocean depths at about 350 points, spaced on the average more than two hundred miles apart. The great depths, the strange life in the deep sea, and the revealing samples of sea floor were exciting to find, but the points were too few to produce a good map of the world ocean bottom. The geography and geology of the oceans—most of our planet—is an accomplishment of the last decades.

DURING THE EIGHTIES A NEW WORLDWIDE location system using precision timing has come into preliminary use. This is still a form of mapping earth by the sky, but it uses neither Antares nor Spica but a brand-new constellation of artificial stars. There will be a dozen or two circling orbiters, satellites with the talent to signal time very crisply from the sky by microwave radio. The present American scheme, a national program managed by the U.S. Air Force, is called the NAVSTAR Global Positioning System (GPS). We persuaded two

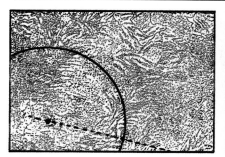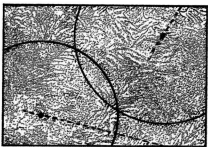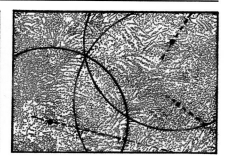

How GPS works. Three satellites send out their time signals: we must be where the three circles cross.

experts to bring their own novel version of a practical commercial GPS receiver to Ellicott's Hill in Natchez, to compare the latest word in timed mapping with what was done there with the transit, the telescope, and the tables of the moons of Jupiter.

The principle of GPS is simple and elegant. The detailed execution is demanding. Begin with the principle, simply described.

We will work out our simplified example in two dimensions, on a flat surface. Suppose that the path of a moving transmitting source across that surface is known at all times. At frequent intervals the transmitter beats on a drum: that is, it signals by a set of sharp coded radio pulses the precise time of a very accurate clock it carries. The radio signal is picked up by a receiver at some unknown location on the surface. The receiver, too, has a fine clock, of comparable accuracy. By comparing the time the signal comes in to the receiver with the time of sending that was announced by the satellite, the receiver knows the travel time of the radio signal. But the satellite also announces its position as each pulse is sent out. The receiver must then lie somewhere on a neat circle of known radius: the center of that circle is the known position of the transmission, and the radius of the circle is just the distance that the radio signal can travel in the known time of delay between the sending of the pulse and its reception. The circle embodies the fact that the receiver has no clue at all to the direction in which the source lies, but only how far away it is.

But now let a second source of the same sort be picked up by the receiver. Now the receiver will learn that it lies on another circle of a different but precisely known size, centered on the announced position of the second source. The receiver must be at some point that is common to the two circles. Two circles can intersect in two points only. A third circle will finish the task; it must intersect with each of the first two, but only at one point will all three circles cross. That is where you are.

The unstated difficulty is easy to grasp. Every microsecond of error in time means something like a thousand feet off in position. To get really good results, timing must be correct down the whole set of links in the chain to a tenth of a microsecond or better. It can be done.

The first task at Ellicott's Hill was to set up the small receiving antenna and radio receiver, all held within a plastic enclosure that sits on a surveyor's tripod. That antenna defined the point on earth that GPS would locate. Cables connected it to two small cases, one that held the precision clock, one the small computer. The satellites we used were invisibly high. There were already about a dozen in operation in the still-unfinished constellation, each one circling around the earth across both poles every twelve hours, 10,000 miles up.

The computer screen soon identified the

```
STATIONS:
  ROVER  :Natchez-corr'd   31d33m44.4400   -91d24m10.8296    67.41
  BASE   :Natchez, Miss    31d33m10.00     -91 23 09.00       .00
COMMENT :Complete single station processing near Natchez, Miss.
BASELINE:      LENGTH=     1946.491
  LOCAL  :  -1630.621.  - .732mE   1060.848  -2.196mN    67.113  -2.865mU
  WGS72  :  -1618.101  - .757mX    537.224   -1.330mY    939.130  -3.350mZ
SITE VEC:
  ROVER  :         .000E          .000N         .000U
  BASE   :         .000E          .000N         .000U
CLOCK   :
  OSCIL  : -57.78    -0.941E-10      -0.412E-13/s
                     0.310E-16/s2   -0.705E-20/s3    0.000E+00/s4
  EPOCH  :    BASE:     .000000sec    ROVER:    .048400sec
DATA SET:
  ROVER  :natchez
  BASE   :
```

four satellites we would use, telling the number of signals timed from each one, and listing the times of all the intervals. We were shown the results of calculating the overall internal consistency of the timing, and were presented with the best averaged results. The complete measurement took about twenty minutes: latitude, longitude, and altitude.

Ellicott's uncertainty was some five to ten miles east and west, but his north-south position was reliable to perhaps one-quarter of a mile, after months of measurements. To calculate longitude he needed to use time measurements, and in pre-electronic days knowing an accurate time at a distance was always difficult.

In our own evening's reconstruction of Andrew Ellicott's work by star sights we managed a north-south position uncertain by a couple of miles, and an east-west one that was within perhaps six or eight miles of our true position, the consequence of a half-minute doubt in timing.

GPS offers its laconic report (noting degree by d, minute by m, and second of arc by s):

Ellicott's Hill by GPS:
    latitude   +31d 33m 44.44s
    longitude −91d 24m 10.8296s

Like most computer programs, this one reports its calculation without rounding off figures that go beyond the goodness of its input data. One arc second of latitude is about one hundred feet, one arc second of longitude a sixth less. (But I surely will not believe GPS

Memo from the GPS computer. The GPS terminal left a printed record of what it had done, given here in part. BASE was the trial map location we told the computer. ROVER was the location it found, including an altitude near 67 meters (it fits the map). There is much technical detail; for example, WGS72 names a particular model for the figure of the earth to which the coordinates are referred.

down to the eighth of an inch of its last decimal point!)

The satellite orbits are not perfectly known; the speed of the radio signals varies very slightly in the upper atmosphere; the shape and size of the smoothed-out earth used to translate from the orbits that feel the earth's overall central pull to a position on the earth's surface are not without error. All these effects mean that the GPS report is correct to fifty or a hundred feet over the whole earth.

I am still bemused by the next number the GPS computer offered that night. Using a fourth satellite, it gave the altitude of the antenna.

The GPS altitude of our spot:
    67.42 meters, or 220 feet above sea level
The altitude from the USGS contour map:
    between 220 and 230 feet.

I have not even checked whether the reference sea level used for GPS is the same as the standard used for the USGS map. The GPS system had put us in all three dimensions at a unique place on (or above) the earth with an uncertainty of only fifty or a hundred feet.

The method is new, still under test; it will get cheaper and better. Radio signals that carry a

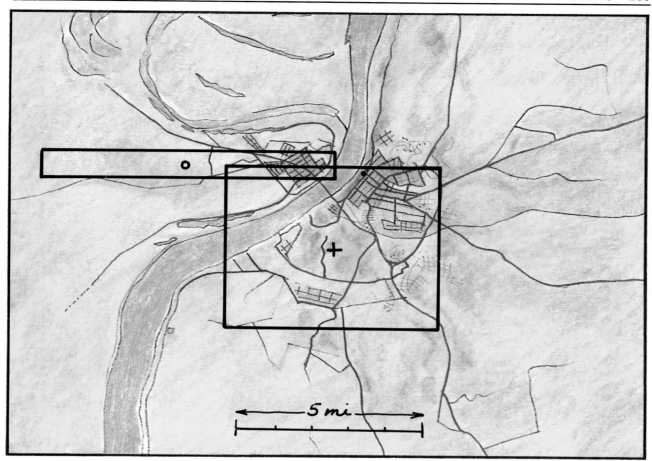

Locations old and new. The map compares estimated reliability of mapping location for Ellicott in 1798, his best value marked with a small circle; our own quick try with old methods (plus a telephone), best value marked with a plus sign; and for the state-of-the-art GPS, best value marked with a dot (to be seen near the top of the larger rectangle). The boxes show the areas within which it is a good bet that locations measured repeatedly by each method would scatter. GPS locations are so precise that the dot itself is large enough to catch all expected GPS results! The base map has its uncertainties, too, which are about the same as those of GPS.

flood of electronic ticks from rigorous clocks sent circling about the earth on artificial stars whose orbits are under our control will before long build up a world map in vast detail, whose framework can be pinned with precision to the earth almost anywhere at all. The careful sightlines that triangulated the nations in the past are probably obsolescent for mapping large distances. That any military action at long range can be based on GPS location of unprec-

edented accuracy is another implication not to be forgotten.

The time-based maps of the future are sure to become so good that we will notice—it has already begun—how the solid rocks of the mapped continents themselves drift very slowly over the round earth. But that opens a story other books will have to tell.

THE TELESCOPE OF GALILEO TURNED THE moon into a place, a landscape in some ways as familiar as our own. The moons of Jupiter were first of the highest intellectual importance as tests for the theory of motion, and then became useful for more than a century in a practical way as the hands of a clock. But those starlets, no bigger than our own moon but a couple of thousand times farther away, could never be seen as landscapes even in twentieth-century telescopes.

The cameras had to fly close to the Galilean moons before Io and her three companions

Three centuries of Jupiter's moons. The first image is a historical account by computer of only the last century of observing and timing eclipses of the innermost moon of Jupiter. Every observation is marked by a + sign. When the point lies on the central zero line, that observed eclipse was exactly as predicted by Dr. Lieske's best tables. The points naturally scatter symmetrically late or early by up to a minute. But there is no bias; the schedule fits.

In the second graph, eclipses of the same moon are followed back all the way to the time of Cassini. True, the scatter is a little larger for the old observers, maybe two minutes. But now we see a drift, a bias. There is no reason for the old observers always to be late; that drift must be in the predictions. The red line shows how the predictions could be modified to eliminate the drift. The points would then be symmetrical at all times; the old data have the power to improve the new.

could be seen as the strange and varied places they are. It was the Pioneer and Voyager space probes during the 1970s that first showed us the faces of those satellites.

Maps are what explorers need, and the space probes were not an exception. Astronomer Jay Lieske of the Jet Propulsion Laboratory had the responsibility for tabulating anew the dance of the main Jovian moons in preparation for the 1979 visits of the Voyager probes, planned to fly close by those strange little worlds. That task led him from the big computers of Pasadena to the crowded archives of the Observatoire de Paris.

"In order to predict where the satellites are," he began, "you have to describe their orbits." Roughly the orbits are circles around which the moons uniformly move. That is not good enough for close-up work. The moons pull a little upon each other as they circle huge Jupiter, and the orbits nod under the off-center effects of Jupiter's somewhat flattened form. Those small irregularities are the heart of the matter. They are an old challenge to the adepts of celestial mechanics, and a good fit is no simple exercise. There are scores of relevant quantities that must be put in to the lengthy calculations.

The dance is still being carefully watched from Paris and elsewhere. The best-timed events are the eclipses of the innermost moon, named Io. Lieske started off his work a decade

ago with a series of about 800 modern eclipse times—Io orbits her planet in about a day and a half—most of them observed during this century. He adjusted his formulas to this set of data until he had a scheme to compute and tabulate the eclipses of the future. It was the best schedule available, and it guided the Voyager probes well.

No table is perfect, not Galileo's, not Cassini's, not Lieske's. No timing observations are perfect, either. If you compute the calculated times of the eclipses over the past hundred years and compare the computed time with each observed time, you expect and you find a certain random scatter. The eclipses were seen sometimes very close to schedule, sometimes a little early, sometimes a little late. The scatter in the observed time was never more than about thirty seconds each way; a perfect theory and error-free observations would have agreed precisely. But the fit was pleasing enough.

Most important, early and late were symmetrical: there was no bias or drift to be seen over the decades. Dr. Lieske's desk computer obligingly displayed the plot of all 800 eclipses. The dots were scattered equally a little early or a little late over the whole century of data.

That symmetry is to be expected if the theory is good, for there is little reason for errors of observation to favor either early times or late ones. A small cumulative error in the schedule can be hidden by the scatter for a short run of

ECLIPSE OBSERVATIONS OF IO

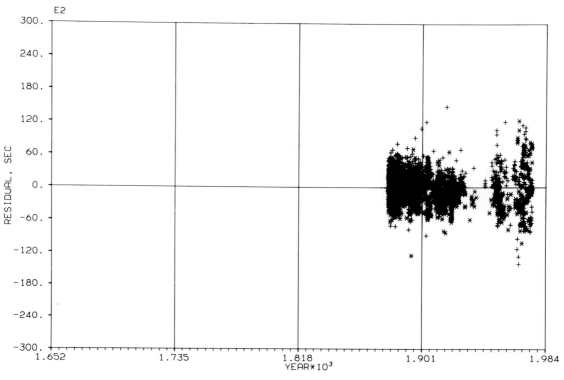

ECLIPSE OBSERVATIONS OF IO

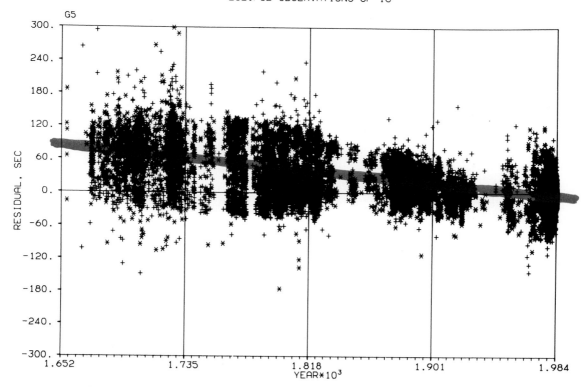

Io

Europa

points, but over a large number of observed eclipses, over many years, it will appear as a slow drift toward early or late times, a visible bias that grows as you go back in time.

The next probe NASA has planned out to the Jupiter system is named Galileo. That spacecraft is intended to orbit back and forth for months in the space near Jupiter, approaching the moving moons to receive small gravitational pushes and pulls from them in a cosmic game of Newtonian billiards. For Galileo's complicated life, a dance among the dancing moons, the moon predictions needed to be as accurate as possible.

Jay Lieske went off to the Paris Observatory to get more data. For a year or two the expert on orbits found himself turned linguist and historian, puzzling over old books in strange notation and over unpublished archives in Paris, looking for long runs of good data on the eclipses of Jupiter's moons. Happily they turned up, many of them compiled long ago by one tireless scholar, misplaced and unpublished in the turmoil of the French Revolution.

Lieske returned to Pasadena with his haul of data from over the centuries and around the world, 8,000 observations of the eclipses of Io, and other moon events as well. There were many good observations by Cassini and by the missionaries Cassini trained in the 1670s, and a few eclipses even before that.

The desk computer again obliged us. It plotted all 8,000 Io eclipse timings on the screen year by year. "Something interesting happens. The modern values did not change, but when we go back to about 1670, it is fairly obvious that there is a trend in the data. Our model is in error."

As you look back in time you see a whole band of 8,000 scattered points from 300 years of data. The band widens a little in the earlier centuries; the local clocks were not so reliable then and the telescope's image not quite so clear. There was therefore a little more scatter in the data, a little more uncertainty in time. But it is a minor difference. The whole band of points bends upward from the line of zero error as it goes back in time, unmistakably drifting. It drifts as a whole, so that almost all of the seventeenth-century eclipses were observed to take place about a minute later than the scheduled time. It is unlikely that these were timing errors by the old observers; how could they all have had clocks that were ahead of time, and never clocks that read slow? What we were seeing was some bias in the system.

The old observers appeared to be selectively biased by about a minute.

That indicates to me that our predictions are in error. We changed the period of Io in our model, and then reanalyzed

Ganymede

Callisto

the same data. You'll notice that the trend has now disappeared. Everything...scatters around...our idealized prediction.

We've had some 62,000 revolutions of Io during those 300 years. In over 300 years that amounts to something on the order of one millisecond per revolution. That's a very small change to make, but over a long period of time it has quite a drastic effect. I have a lot of confidence now in our ability to predict the moon's positions on into the future.

Thanks to the observers of the seventeenth century, Galileo will more confidently approach one satellite after another in its rounds. The Galileo spacecraft had been scheduled for a shuttle launch in the spring of 1986; it is not yet rescheduled. The gravitational motions of the distant moons are more predictable than the misfortunes of our own intricate endeavors.

The direct aid that Cassini's work could give to the Galileo mission across three centuries is an eloquent statement of continuity in science. A good measurement has lasting worth. The circling of the Jovian moons poses a problem that is shared by the astronomers of the seventeenth century with the space scientists of today, in spite of everything that separates them.

GALILEO'S MOONS

OUR MOON IS A PLACE SEEN FIRST by Galileo. But the Jovian moons became real places, if strange ones, only when spacecraft carried video cameras nearby in 1979. They are here in enhanced color, images to scale, and in order of increasing distance from the planet. Only Europa is smaller than our own moon, and only by a little.

The Music Mountains

AGREED, THEN, THAT MAPPING IS OF THE highest importance. Maps are surely important to the mind, for as human beings we always seek to know where we are. They are decisive in practical affairs, whether we think of a simple map of the street corner, or a grand map of the whole earth or the solar system.

I believe they hold more important meaning still. They underlie a grasp of science as a whole; we often make visual representations, surely to be called maps, of forms that do not exist at all in the space where we dwell. I am not talking about the geography of fantasy, like

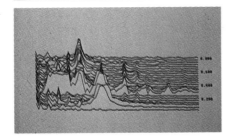

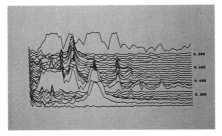

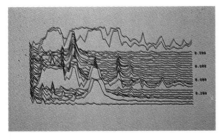

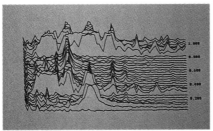

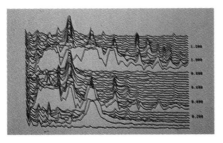

The music mountains caught in a few still images from the moving map of music. Peaks are loud; low tones are mapped to the left; time passes smoothly from one profile to the next.

a map of the Land of Oz. Such a map is still recognizably geographic. There are the brick roads and the dark woods, even the directions of east and west, though all in fiction. The maps I mean have left ordinary geometrical spaces behind.

We found a computer group that had been working upon visual representations of performed music. Their map of music is much more representational than a musical score, which is a kind of compact coded sketch; the score only instructs, representing what we hear only at considerable remove.

We watched their map of a piece of music unfold. This map refers to an entirely imaginary space, but it is no fantasy. It is as matter-of-fact and as faithful to the actual performance as a good road map is to the state highways.

The map they made of a rhythmic popular tune from India is a moving map; the video screen presents a wide overall view of the contours of a strange dancing mountain landscape, as though we were watching from high above a steadily unrolling scroll that bore an animated map. The landforms move in perfect time to the music.

Anyone who watched the moving image pass as the music plays is soon able to pick out at least one of the rules of this mapping. The peaks are associated with loud tones; in the valleys, there is not much sound. Less obviously, the music is sorted by computer calculation

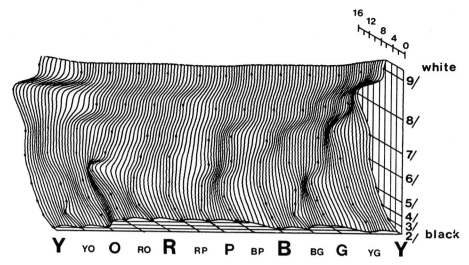

Y YO O RO R RP P BP B BG G YG Y

A map of color space. The sculptured form presents a theory of human color vision (by no means the final word). The colors are mapped in three dimensions. Along one of the horizontal directions all the hues are mapped: orange, red, purple...on back to orange again. At right angles to that direction runs a scale of light and dark, every degree of gray from black to white. The vertical height above the base is a measure of the color saturation of a pigment, independent of its dark or light quality. Every surface color we see has some place within the solid. The rolling plateau maps the most saturated pigments available.

according to its instantaneous pitch, much as the ear can tease out the several notes that make a chord. The treble is drawn to the right; the low notes, toward the bass, are mapped at the left side. The computer draws such a profile by a tedious analysis of the taped music by loudness and pitch, working on a sample that is taken many times a second. The process repeats over and over, sample after sample, as long as the music continues. Then all those split-second profiles, thousands of them, are exposed in order to a video camera, and the whole series is screened rapidly as a motion picture to run in time with the music on the original tape.

The diverting landscape of the Music Mountains is something that was constructed by reasonable rules from a piece of music. It was not literally in the music. The music certainly has no left and right, no literal peak and valley. Music is not in any geometrical space at all. But given that conventional structure, the map represented the music faithfully, all its changes, all its tones.

Such maps are persuasive to the mind. They crowd the literature of science and technology, almost always in much more austere and utilitarian form than ours, much simpler and not in motion. They are usually called graphs. Those are found in almost every book and paper in science and technology, commonly even in the business pages of the daily newspaper. But they

are all of the class of maps, maps in space—though often drawn only in the flat—but maps typically representing what was never and cannot ever be found in space.

Ideas commonly seem to dwell in space. We often speak of one idea as being behind another, of two notions being close, or of two statements requiring a connection based on some concept in between. Scientific theories quite generally resemble maps, even when they are complex. Theories almost always concede that they have some built-in limitations and some approximations. They often bear some marks of original and now outworn purposes. Sometimes theories shift, as suddenly as once the mapped coast of the Sun King's France shrank before his eyes. Sometimes they endure firmly over thousands of years, like our recognition of the roundness of the solid earth. We expect them to improve as did the maps of the world; we hope for exciting new discoveries from time to time, and we can count upon steady cleansing and incremental improvement in detail. But one thing we ought not to expect: with our theories, our maps of the processes of the universe, we should not expect to make simple pictures of what was never visual in the first place.

But we will always try.

The Norris Geyser basin on a summer afternoon. The boardwalks lead visitors across the flats amidst little hot pools, flowing streams of hot water, and proper geysers, intermittent hot fountains that burst on every scale, from inches to many stories high. They mark the shifting endpoints of the intricate underground plumbing that brought groundwater close to the source of heat below and up again.

4

CLUES

Clouds of hissing steam and the sound of boiling water fill this shallow depression surrounded by the wild hills, as surely they have done, irregularly waxing and waning, over thousands of years; these percolators and boilers owe nothing to the work of human hands.I had come to the Norris Geyser Basin, one of a dozen areas within Yellowstone Park, where altogether there can be found half of all the active geysers in the world.

It takes only a little experience in the kitchen and some reflection to gain the clear idea that this boiling water and steaming surface must somewhere have come from a hot caldron, with the heat to keep it going for a very long time. But you will not find those anywhere in the woodlands or on the hilltops. From the location I can safely infer that those essentials must be somewhere under the surface, hidden from view.

That is an everyday conclusion almost anyone can form. Yet it is the primary explanation of these curious wonders, an explanation the geologists have confirmed with all the richness of detail they add. The caldron turns out to be no empty cavern, but a labyrinth of natural plumbing, an eroding maze of joints and cracks within near-surface rock. The heat is provided by the rocks deep underground, less well mapped by the geologists, but certainly only one legacy of the long volcanic history of this hot spot of earth, where once upon a time tides of lava and storms of ash flooded out to leave the landscape as their residue.

Written everywhere in the landscapes of the world and under its oceans are more subtle hints and clues. They allow us to raise still harder questions, not only of the present, but of the deep past. We can read there natural dramas in many acts, events more remarkable than any human eye has ever witnessed. The evidence of geology, like this geyser basin, is often on a scale that requires no laboratory instruments, but can be ordered into a narrative account of what happened, much as the classical detective of fiction pieces together clues that anyone might see and come to grasp. It is this quality of field geology that is exciting; in this science, participation is not yet entirely fenced off behind a barrier of mathematics and of complex instrumentation.

THE FIELD GEOLOGISTS CONFRONT THE EVIDENCE

RICHARD YURETICH PROVIDES SCALE for a standing fossil tree, and his hammer and a lens cap lend scale to its details. The textures of the deposited rocks are characteristic. One embedded fossil tree root is to be seen near the hammer, and the growth rings across the fossil stump are eloquent. The same standing trees were photographed during the 1890s by earlier inquirers. The field sketches made by W. H. Holmes around 1878 were the basis of his first detailed record of the fossil forest layers on Amethyst Mountain nearby.

These big fossil trees are witness to the forest ecology of that ancient era, a time when the sequoia and its kin were the glory of Wyoming valleys. Fossil leaves and needles and forest litter of all kinds are common in these layers, but there are few or no animal fossils. Any creatures able to move fled the volcanic outpourings.

1986

Old photos, 1890s.

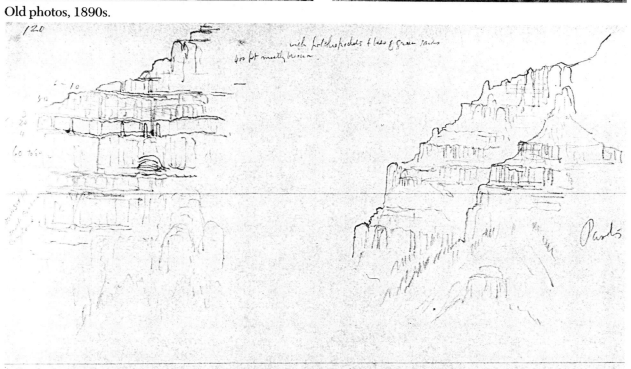

Field sketches, about 1879.

RICHARD YURETICH, A GEOLOGIST AT THE University of Massachusetts, Amherst, has walked the valleys and climbed the slopes of Yellowstone Park for some years now. He guided us in clambering up one rocky slope above the wide valley of the Lamar River, in one of the wilder corners of the big park where the bison were quietly grazing.

A good geologist does not ignore even the loose stones at his feet. "As we came up this slope I found some unusual stones. They look like rocks, they feel like rocks, but if you look closely you will see a structure and a texture that resembles wood. That is the first clue that on top of this hill, from which these stones rolled down, something truly unusual is to be found."

There were early tall tales in plenty, but more than a hundred years ago, the geological artist W. H. Holmes made a serious study of Specimen Ridge and its Fossil Forests. What he did was aim his telescope at the rock face from down in the valley. He drew the layers he saw and the trees, standing and fallen, he saw in them. Embedded within the different layers were upright stumps and trunks that appeared to be the remnants of patches of forest growing in place, buried by what we now know as volcanic activity.

On the slope here and there still stand fossil tree stumps of rock. Some of them are exquisitely preserved and not hard to reach. We can actually see the individual growth rings, which give us an idea of the age of the tree at the time that it died. Richard made a quick count: over 200 growth rings. That tree once grew quietly for some 200 years until it was buried in place. Near it is another standing trunk, and at its base you can see a swelling root. But when he hammered the root, it too rang like solid rock. The fossil tree root is embedded in another fine-grained rock, a sandstone. Directly above is a discolored rock whose texture is that of an old layer of soil, and a foot or two above that a darker rock of coarse irregular texture.

That tree once grew in place in sandy ground. A richer soil cover slowly formed in the woods. During the next volcanic event, which produced the mud flow that remains as the coarse conglomerate, the tree was buried, overwhelmed by the sudden tumbling mud. We have no exact parallel for the burial of a forest on the scale we see in the fossil forest here. But the eruption of Mount Saint Helens in May of 1980 provided us with an extremely useful analogy.

First of all, that explosive eruption spread a tremendous amount of fine-grained ash which settled over a wide area in a neat layer. That layer is equivalent to the fine-grained ash-bearing sandstones that we see in the layers of Specimen Ridge.

Another good match is to the mud flows that the 1980 eruption spawned out of huge floods

Witnesses in plenty saw the events around the eruption of Mount Saint Helens in the Cascades beginning in May 1980. These photos evoke the ancient events of the Lamar Valley. The line of green growing trees is caught at the moment of arrival of the inundating ash and dust carried in the hot gases of the volcano. The desolate modern forest, many dead trees still standing, makes plain the consequences of a blowdown, and suggests an episode in the narrative of the layers on Specimen Ridge.

A chocolate layer cake is a model of the layering of sedimentary rocks. Of course, its artfulness idealizes the complex reality of a piece of landscape. But the inference to be drawn from the removal of a slice of cake is still basic. Layers that can be matched and followed from one side of a valley to the other must once have been more or less uninterrupted. The order of placement is clear.

of melted snow on the mountain slopes. The wet mud cascaded down the Tootle River Valley. When the flows reached the flatter parts of the valley, they spread out to cover extensive regions of the forest that was growing there. By now the flow deposited by Mount Saint Helens has been reworked by organisms and new plant growth; a good soil is forming, in much the same way that fossil soils developed upon the ancient ash and sand layers here on Specimen Ridge and Amethyst Mountain.

YURETICH: Looking at the arrangement of these fossil trees, we could see that these two trees are embedded in lower rock units than are the ones up above us. These are therefore older. One of the basic principles of geology is that when you have a series of horizontal strata or sedimentary rocks as we do here, the ones that are lower down in the sequence came earlier than the ones that are younger and hence higher up.

A chocolate layer cake offers a cheerfully instructive model of such basic geological interpretation. Entire, it is a tempting icing-covered cake. But no one can enjoy—or understand—the cake without going below the surface cover. A cut or two will do that nicely. Now the cut layer cake reveals the history of its inner construction. Flat layer was placed upon layer. The deep valley I had cut out, the opening anyone makes to slice a cake, did not exist when the cake was made. The missing section

has gone elsewhere. But surely the layers were once complete, smoothly crossing the place of the missing slice one above the other. Only later a knife cut at either side revealed the layers within.

Even if the knife had entered at a slant, so that instead of a steep cliff face we had made a gentler slope, it would still follow that the lowest layer of the hillside would have been set in place the earliest. Of course, to be quite sure of that, we would have to be convinced that the substance we examined was an integral part of the layer of the cake. So to that end we take a small sample (that works for the geologist, too) to make sure of identification: chocolate—or tuffaceous sandstone.

On Specimen Ridge there is a thick series of repeated flat layers with standing fossil trunks, stumps, roots, and fallen logs buried in one layer out of every two or three. Once those layers were flatlands; there was no Lamar River and no valley opening. The flat layers stretched largely unbroken across the present valley many hundreds of feet above its floor, where now only birds and helicopters fly. Those layers in succession are composed of the very rock types expected from ash and mud flow, soil, new mud flow, and so on.

Note that if a tree grew with its roots in one layer, it did not live at the same time as a tree that is rooted in a higher layer. Today the standing fossil trunks look as though they had all grown at one time, but stood on a slope. That was not so when the ground was flat, before the

A layer cake of forests was what Mr. Holmes drew
from his early studies. The section he drew here is
not literal; he could not see inside the rocks. But it
is an imaginative reconstruction of what is buried
within the mountain judged from the exposures he
saw. Today we agree broadly with what he inferred
long ago.

East Fork.

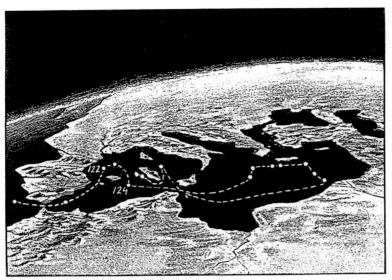

The route of leg 13 of the Deep Sea Drilling Project of the international group, Joint Oceanographic Institutions for Deep Earth Sampling, under the sponsorship of the National Science Foundation of the United States. Each of the specific voyages of the DSDP was called a leg. Leg 13 was the Mediterranean study that began with the search for mysterious reflector M, and the older history of the whole sea. That voyage down the length and breadth of the sea took two months in the summer of 1970. Each drill site is numbered; it was site 122 that yielded the first bucketful of implausible gravel, and site 124 the cores from the desert lakeshore described in the text.

cake was cut or the valley formed. The lower trees grew in a forest that had died out long before the higher one was even seeded.

The stands of fossil trees constitute a sequence of successive mature forests, one buried on top of the other. An eruption from the ancient volcanos whose dead cones stand only miles away from here started a rain of ash and a flow of mud. The forest that was then growing would die, perhaps in a matter of hours or days. After many years soil would form, and trees could once again grow among the old half-covered trees, mostly fallen and decayed. Centuries would pass, and the second forest would mature, a new tall forest of large trees. Suddenly another volcanic event would kill that forest in its turn, and in due time another one would arise and flourish, to be brought down in time by yet another eruption, and so on, forest by forest, until the top of Amethyst Mountain is reached, after hundreds of feet of volcano-caused deposits.

Richard Yuretich counted about a dozen layers of fossil forest along the route he climbed that day. Of course a few hundred square miles of mountain lands are not as simple as a layer cake. Here and there are bound to be openings in any forest, the places of streams, ponds, fires, poor soil. So the layers are not uniform, one just like the other, like the cake. There are slumps and breaks as well. Sometimes a whole patch of standing trees in place might even slide as one downhill under some mud flow, to confuse the record. But if you put together all the fossil forests now partially exposed, you may roughly count two dozen or more nearby forests that grew, one almost on top of another, each to vanish in turn. The whole episode took 10,000 or 20,000 years, one stage during an era far longer into the past than that.

It is an outrageously tall tale—two dozen forests in stone one above the other—almost as bad as the sly old jokes of the petrified song from the beaks of the petrified birds on the petrified trees! But this evidence is so direct, so visible, so coherent, that we can hardly doubt it. It takes little instrumentation, little geochemistry or radioactivity, little math, to work the story out; the geologists of a hundred years ago found the main outlines of the drama during the first seasons they spent in the valley. If you pay attention to the rocks, you can check it yourself on an enjoyable trip to Yellowstone Park. It is one of the surest ways to gather some appreciation of the depth of geologic time, and the strangeness of the events that can happen in all its long span.

Some Boots!

A PAIR OF GOOD BOOTS AND A SMALL HAMMER are manifestly adequate equipment for studying the slopes and cliffs of Yellowstone. But most of the earth is not in an open park, but is

THE SHIP

BOOTS AND HAMMER HAD TO TURN into a great ship to give geologists access to the rocks beneath the sea. The drillship *Glomar Challenger* was built for the Global Marine Corporation on the basis of experience in seeking oil underseas. Turned to research, for fifteen years she drilled below the seven seas with her powerful diesels, a seagoing oil derrick that avoided oil. The drill stem goes right down amidships through an opening in the hull; the photo looks down into the blue moon pool on a sunny day. The deck is the rig floor, and there the roughnecks perform their unending feats of rigging and assembly, manipulating steel by the hundred ton on board the heaving ship as deftly as though they were on the plains of east Texas. The weight of these forgings is manifest; this is a tool on a massive scale. The business end is the drill bit, its hard teeth always at work a mile or two below the ship. At the end of the drill stem it grinds its way through rock hard and soft; during any short distance it can be managed so as to leave behind an undisturbed central core, to be reclaimed as sample.

THE FIRST CHALLENGER

IN EVERY LIBRARY WITH A CONCERN FOR oceanography fifteen feet of tall shelves are weighed down with one big set of ponderous matching volumes. They carry the diverse cargo of knowledge brought back and amply published as the *Report of the Scientific Results of the Exploring Voyage of HMS Challenger during the years 1873–1876* (50 vols. London, Edinburgh, and Dublin, 1885–95).

The *Challenger* was a 200-foot wooden steam corvette of the proud Royal Navy, three-masted, square-rigged, and powerfully engined as well. Two of her seventeen guns remained; the removal made room for a chemistry lab in one gun bay and a working darkroom, and even the captain's quarters had been reduced to accommodate the director of her half-dozen scientists at sea. The ship's crowded company was about 240, including naval officers and men, the scientists, and an official artist. She sailed a track three times the distance around the world, touching many a lonely island and all the continents, except that she prudently stayed clear of the Antarctic barrier during her two weeks amidst the pack ice.

Surveys for the new submarine telegraph cables had in the 1850s aroused interest in the ocean depths, and during the 1860s a number of pioneer voyages had been made to sound the waters, to take samples, to look for life in the deep and the dark. But the *Challenger*'s was the first worldwide voyage of oceanographic research, well-supported and brilliantly published in high Victorian style. Close to a hundred specialists from many lands contributed to the final reports drawn from the take of nets and dredges, the thermometers and current logs and seawater samples, and the soundings made at 350 stations at sea. The colorful large lithographic plates of microscopic radiolaria and big starfish, of pelagic jellyfish and foraminifera from the bottom ooze, of manganese nod-ules and luminous fish and even of the cormorants of the islands are still exceptional as natural history and as illustration.

It was natural enough to name the first ambitious scientific drillship after HMS *Challenger*, and it is touching that the shuttle orbiter lost in the launch tragedy of 1986 was also a namesake of the tall ship that made so happy a voyage for science more than a century ago.

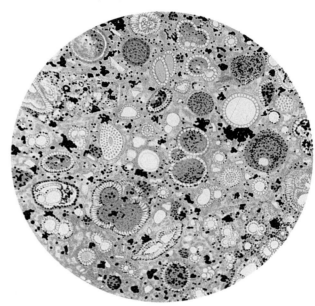

The circular field is a microscopic view of bottom ooze from a spot in the South Pacific, a mile and a half deep. The plate to the right shows a colorful living radiolarian, about the size of the dot of an i, surrounded by the artist's arrangement of skeletal fragments of related forms.

H.Haeckel and A.Giltsch.Del.

K.Giltsch,Jena,Lithogr.

1 - 8 LITHOCIRCUS, 9 - 17 DENDROCIRCUS.

CORES

THE FIELD GEOLOGIST AS HE WALKS the hills hammers off chunks of interesting rock to carry back to the lab. The basement of every old geology department is lined with heavy wooden shelves that hold the specimens of the past. But drillship geologists face the same task differently.

The ship is able to compensate for heave and drift, to keep itself in place directly above where it is drilling on the sea floor. A mile under the sea surface and on down into the bottom sediments a big hollow drill bit is kept rotating night and day at the end of the twisting drill stem. The hard bit steadily grinds away at the rock. Fluid is circulated through the drill string to wash away the cuttings. Once a sample is wanted, it is time to bring out the undersea geologist's hammer.

First the drill stem is disconnected from its driving gear at the derrick. Down the long hollow stem is sent a thinner steel tube about thirty feet long, the core barrel, lined with a thin plastic tube. Once the barrel is seated at the bottom, the heavy drill stem is reconnected, and the bit begins to turn again. The washing fluids are turned off while the bit drills another thirty feet deeper. The lined core barrel should then be filled with an undisturbed cylinder of whatever material is there where the hollow bit has just ground its way. Again rotation stops,

the drill string is disconnected, and a fishline of steel cable is sent down to hook the top of the core barrel to bring it up. The process continues, core on top of core; or steady drilling can be resumed to greater depth. There is plenty to go wrong, working at the end of a long string of pipe in a deep well hole that penetrates all sorts of rock a mile beneath the sea.

But if nothing goes wrong, the core is a specimen bound to excite the envy of the man with a hammer. It is a complete sample from far beneath the sea of thirty feet of deposits, hard or soft, ordered in time, the deepest layer the first to be laid down. The roughnecks pull the filled liner out of the core barrel once they bring it to the deck of the rig. There it is sliced into six sections, labeled, and examined externally in the shipboard labs. Finally, each section is sliced lengthwise; half is stored for the archives, and half taken for detailed study aboard ship.

The core library at Lamont-Doherty Geological Observatory has on its racks the archived half of 20,000 cores or more from beneath all the oceans. So much organic matter is present in the cores that it is prudent to keep this library without books well refrigerated. The earth's past is shelved here in a significant sense, its record written in the many languages of nature.

Within the sea floor.   The drill stem enters.   A cutaway view.

The two co-chief scientists of leg 13 examine a core. The shipboard picture shows Bill Ryan, Ken Hsü in his hard hat, and drilling superintendent Travis Rayborn of Global Marine.

The shipboard paleontologist of leg 13 samples some muds. Maria Cita of Milan is photographed aboard ship.

submerged at the bottom of the deep salt sea. Look at any map of the world. It is mostly blue, mostly ocean. Ours is a watery world on the surface, but underneath the ocean everywhere is solid rock, the planet.

You cannot study geology thoroughly on this earth without looking under the ocean. But no geologist has ever walked in his sturdy boots with a light hammer in his hand across the bottom of the ocean looking for samples. We could have only incomplete knowledge of the world's geology until we found adequate means to sample the rocks beneath the ocean ooze. Those means have come about only with people still at work today, people who went on the ships to drill down into the rock below the bottom of the ocean, to sample for the first time most of the surface of the earth.

Some boots, some hammer! Instead of those homely aids, the indispensable tool of the modern undersea geologist is more like the *Glomar Challenger*, a specialized four-hundred-foot oceangoing drillship. It can work with miles of heavy drill string, five-inch steel tubing that it can hoist high and turn steadily from a heavy derrick like that over an oil well, to drill into the layers beneath the sea bottom. At sea it was home to a hard-working population of about seventy-five men and women, ship's crew, oilfield roughnecks who worked the drill rig, and a variety of shipboard scientists and technicians. The ship cost a million dollars a month

to run. She was put to basic geological research by the Deep Sea Drilling Project of the National Science Foundation between 1968 and 1983.

THE STORY OF THIS PARTICULAR TRIUMPH OF the Deep Sea Drilling Project is the story of one outrageous idea. The idea came to two exhausted young geologists in the small hours of one troubled summer night at sea. On the evidence of a handful of sorted gravels just fished out from the rocks beneath the sea, they set out the "very very crazy hypothesis" that once the whole deep Mediterranean Sea had been a dry desert basin.

The two were William B. F. Ryan, now senior research scientist at the Lamont-Doherty Geological Observatory of Columbia University, and Kenneth J. Hsü, now professor of geology at the Swiss Federal Institute of Technology, Zurich. But in August 1970 they were two young geologists who had to carry for two months at sea the heavy responsibility for direction of that multimillion-dollar drillship. When these two were chosen their critics grumbled that the unique and expensive ship had been put into the hands of "a student and an amateur."

The drill ship operates twenty-four hours a day, so you need several shifts. There are two co-chief scientists; one is awake and directing the operation at all times. Ken Hsü was a sedi-

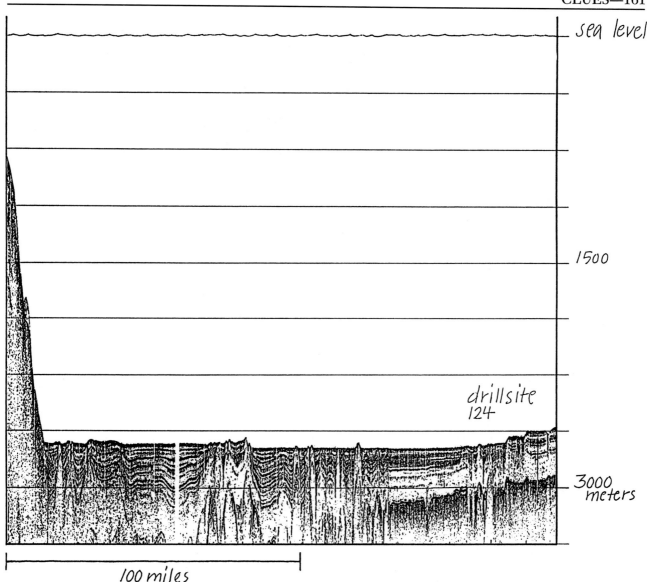

sea level

1500

drillsite
124

3000
meters

100 miles

Echoes from the seafloor map the bottom ooze and far below it. The sea surface is at the top; that flat bottom is two miles deep. At the left edge is the steep drop from the Algerian margin of Africa. The map continues 200 miles north, toward drill site 124 on a mid-sea rise. The strong reflection from the hard M-layer is best seen in the right-hand third of the section; it is that lowest black band a full thousand feet below the bottom of the sea. (Vertical distances are exaggerated 100 times.)

mentologist; he could look at a layered sample from the ocean floor and deduce the environment in which the layers had formed. Ryan was expert in the larger structure of the region, its geophysics. He could direct the ship to good drill sites. They were complementary in personality, too, as they themselves make clear, recalling the early course of "a friendship born at sea."

HSÜ: I was the impatient type. I always wanted to get down through the sediments to the hard basement rock, the end of drilling. Bill was rather deliberate. He wanted to get as much information as he could. Thanks to him we have much of the data we have from Leg 13.

RYAN: I was more systematic—wished to probe gradually down through the sedi-

Hardened limestone oozes.

Ocean basalt.

The unsorted gravels, slightly enlarged.

ment layer, to recover materials consistently, figuring that the real history would be deduced over years and years of study of the cores.

HSÜ: Well, my style was to drill as fast as possible. So we actually drilled more holes on our leg than in any other in the thirteen years of the ship's operation, to find all the different types of sediment we were looking for. You can see it in the operation record: when he was asleep, we drilled fast, and when I was asleep we were getting lots of samples.

RYAN: Ken was always in a hurry. If the material that came up wasn't interesting he would skip, to get deeper faster. He was out for the dramatic answer. That's what the Med gave us—a quick, exciting answer.

They had brought a problem into the Mediterranean with the ship. A thousand feet below the bottom, there lay buried a curiously hard, sound-reflecting layer called the M-layer, M for Mediterranean. It followed the form of the bottom of the sea above it very closely. The layer was known only by the strong echo it returned after delay, an echo from far below the sea floor. There it was, well below softer bottom sediments, a hard layer lying above other softer and deeper sediments, clearly seen in the seismic echo profile across the entire sea from end to end. It was everywhere, pan-Mediterranean. Ryan had done his doctoral thesis on that M-layer, and now he was anxious to drill into it. What could it be?

Crystals of gypsum.

Stressed marine creatures.

M was elusive. At their first drill site within the Med, not far from Gibraltar, it was missing. On the second site, they hit it.

"We hit it," Ryan recalled, "and it hit us back. The drillstring started to bounce and jump around. We knew that we had hit something hard. Then the drilling supervisor said, 'We're stuck. I've stopped moving.' And we were locked in the seabed."

They were lucky. The drill bit after a while did come unstuck, but all that they could salvage from the drill washings as a sample was a lot of gravel. Those bits of gravel had plugged the rolling drill teeth; their rig never could penetrate a deposit of loose gravel, something fortunately very rare in the undersea layers.

The two young geologists had been directing the ship and its expert teams for ten days. All they had to show for it was sleepless nights and hard work. They were in trouble. Their first real haul had just come up from the mysterious M-layer beneath the bottom of the Mediterranean Sea, and what they had was only a few bucketfuls of unusual loose gravel. Very quickly, with the help of the paleontologist of their team, they fished out of those disappointing gravels a wild and wonderful claim: the whole blue Mediterranean Sea, they asserted, had once been a hot, dry, deep salty desert basin, the deepest and widest desert basin the world had ever known.

THEY LENT ME A PINT OF THE VERY FIRST sample, fetched up to my dining room table from the M-layer far below the bottom of the

The entire sorted set.

Mediterranean Sea. The gravel had been stored ever since 1970 at the core library at Lamont, near all the other cores taken by the *Glomar Challenger*. Anyone can sort it by eye, just as Bill Ryan did the night that it first came up. Gravel is only pea-sized fragments of rock, tumbled together by water and gravity.

No geologist, I sorted it entirely by visual appearance, like sorting a pack of cards into suits, the dark bits from the light ones, just as they came to the hand. The dark pile was all of a kind, even under a hand magnifier. All the dark fragments are of the same rock, as far as I could judge.

But the white pile, especially under the magnifier, was different. I could at once see in it two distinct kinds of pieces. One kind stood out: crystalline, shiny, flat, almost transparent crystals. I divided the pile into the transparent crystals and the rest, opaque chalky pieces not at all like the crystals.

All those crystals look about the same. But the chalky stuff invited more sorting. Under the lens I saw some wonderful forms in there that just didn't look like the rest. They were tiny seashells, the remains of little sea creatures. I don't know what they are, but certainly they are no pieces of chalky rock, though they share the color. I had myself then partitioned the gravel from the M-layer into four distinct piles. I saw almost nothing else in the gravel,

perhaps one or two uncertain and puzzling bits out of hundreds.

That simple four-part gravel carried an astonishing message to the knowing geologists.

Ocean gravels:

Gravels themselves really do not form on the seafloor, though plenty of gravel is found there, gravel transported from land by unusual currents. The drill site was only a hundred miles or so at sea. Could these be gravels carried down from Spain? They were much too simple; that four-part recipe was unprecedented. Continental gravel is typically rich in many types of rock. Almost any little stream on land will show a richer mix, a whole assembly of ingredients. In particular where are the quartzes, so common on land, yet quite absent from this sample? This was gravel, but it was an oceanic gravel. It was not from the mainland; no one had seen anything like it before.

Common subsea rocks:

The dark rock was ocean-bottom stuff, the volcanic material common on seafloors. They knew that they had drilled near an underwater volcano. The opaque chalky bits were a limestone ooze, equally common on the seafloor. But this ooze had hardened to rock. One process to harden such a limey ooze is long drying.

The giant Hoover dam spans the Colorado River, carrying the highway on top of its massive arched concrete wall as high as a skyscraper. The filled reservoir lake seen behind the dam seems of commensurate size.

Marine life under stress:

Maria Cita, now professor of paleontology in the University of Milan, was the shipboard paleontologist who shared in that quick unriddling. The most rewarding form of paleontology for the study of cores is micropaleontology, its interests and expertise centered in the manifold tiny organisms of marine and fresh waters. They leave their remains everywhere, in countless numbers. You can expect them to show up within any pintful of a context that might have supported life or entombed the dead; if you wait for rare showy finds, like whale jaws or dinosaur bones, you will be lucky to offer an interpretation of a core once in a blue moon.

Cita recognized the little organisms in the gravel. These tiny shells, millimeter-sized clamshells and spirals, looked very much as though they were dwarfed fauna, small because they had grown up under environmental stress, survived in marginal circumstances. Some dark pebbles had shells attached to them; the animals had grown in place. That strengthened

the view that the creatures had not thrived in life. Sometimes the process of transport can sift small shells out of a more normal population of shells, to present the misleading look of a dwarfed fauna. But these shells had not been carried from afar.

Crystalline gypsum:

The unexpected crystals were first identified by Cita. They were gypsum (given the beautiful name of selenite when in crystalline form). It is a common mineral, always present in solution in seawater. If as you read you are in a room that has plastered walls, you are surrounded by solid gypsum, though not visibly crystalline in form. The plasterer knows very well that gypsum has a special relationship with water. Such bright crystals of the mineral gypsum are commonly formed by long evaporation of seawater on a baking-hot shore.

Gibraltar is the right spot for a dam, though one too large for human engineering.

The four-part message of the gravel was eloquent. A stream down the black rocks of an old seafloor volcano had washed and tumbled these bits of rock along an arid flat to the shore, where evaporation was strong, and the little marine animals endured a short, stressed life. The materials of the gravel were all oceanic; they belonged there on some old seafloor. But the conditions of their presence could not be those of the deep sea. The gravels spoke in one voice of a sunstruck desert shoreline.

Could that buried M-layer be the mark of a hard desert surface formed when long ago the whole sea went dry? The three young scientists were impressed with the aptness of their utterly unexpected conclusions, formed during brief hours, or at most a day or two, of shipboard debate and wonder. But how could you dry up a whole ocean to explain a bucketful of gravel? It was preposterous.

DRYING UP A WIDE BLUE OCEAN SOUNDS LIKE an exploit for Superman. Is there any sense at all in such an idea? How can a sea dry up?

Readers may recall a wonderful construction of 1936, the Hoover Dam. It was really big.

The taxpayers knew that, because it cost a fortune. The engineers knew it, and they were proud of their design and their arrangements. The construction workers worked for a long time, the men always tiny amidst the great walls. Nearly a hundred men were killed during the ambitious project; their names are carved there.

Now change your point of view to a wider one. Look down on the big dam from high in the air. Lake Mead behind the dam is certainly conspicuous; you cannot miss it. But the dam? First you fly down one arm of the enormous lake to look for it, then down another, and maybe one more still: the dam is hard to find. And from orbit the dam that we thought was so big is all but invisible.

The cleverness of the Hoover Dam, its essence, is not in how big it is, but in how small it is. The engineers had found the place where they could make a huge lake simply by closing quite a small valve. Even if the valve was the biggest thing humans could build, the lake which is its purpose is on the altogether greater scale of the wide landscape. Yet a structure so small that it is hard to find in an overall

view controls the whole of the lake, thousands of times greater in bulk than the dam. That is what counts.

Look at the Mediterranean Sea in orbital photograph, or on a good map. At its Atlantic end, there is a narrow gate between Europe and Africa at Gibraltar, a place where a small stopper might make a great difference. You will not find any such constriction in the Atlantic or the Pacific or even the Gulf of Mexico. Gibraltar is a strategic spot for a dam.

Too small to see from orbit, the Hoover Dam is certainly there at the end of one or another of the arms and bays of Lake Mead. But which one? Lake Mead's sheet of water is very much larger than what we saw from the dam. The lake spans most of the hundred-mile width of view of the Landsat camera 600 miles up, and the wall of the dam that loomed so big is now hard to find. The scale of the lake reveals the true nature of a dam: as small as it can be.

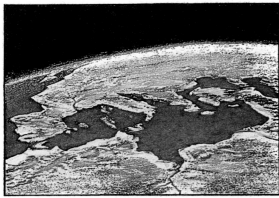

The sea dries up, drawn in four stages, a swift sequence over a few thousand years. A shifting set of salt lakes would stay behind.

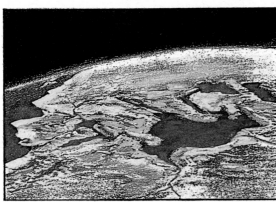

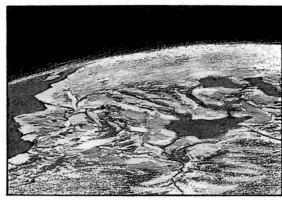

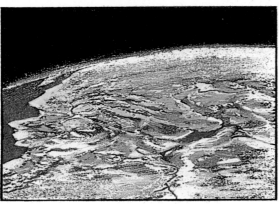

What would happen if the Straits of Gibraltar were replaced by land, and that narrow sea passage between the Pillars of Hercules were sealed tight? Such a dam would have to be a mile high and fifteen or twenty miles long and broad, but imagine it. Strange to say, the whole wide Mediterranean Sea would then dry up within a few thousand years. The rains and the big rivers that feed the sea are simply not enough to keep it from drying out. Evaporation under the hot sun beats them easily. The sea remains watery today only because it receives huge currents of ocean water from the open Atlantic. The currents through the Straits of Gibraltar carry as much water as the flow of a thousand Niles. Salt water both enters and leaves the Mediterranean Sea at Gibraltar. The U-boat captains knew that in World War II, and drifted silently with the currents both in and out through the straits, by choosing the right depth. Less water flows out than enters. The net balance makes up the losses of evaporation. (Suez and the Dardanelles have quite unimportant flows.).

Closing Gibraltar with a dam is entirely beyond human budgets and human abilities. But the land barrier that could do it is still not a large piece of countryside compared to the vast continents of Europe and Africa which would form it. Those two land masses shift and heave very slowly to and fro as they drift over the face of the earth. It is plain that they could open the straits quite wide, or close the straits tightly,

A driveway puddle dries up.

A core reader in front of some stacks of the library of cores, at the Lamont-Doherty Geophysical Observatory of Columbia University, in Palisades, New York.

so to speak without even trying, by chance motions of a few tens of miles out of the thousands of miles they drift over the span of time.

The Mediterranean Sea is vulnerable, unlike the present Atlantic. Its drying up is a real physical possibility, and not just plain absurd.

If a whole ocean dries up, what can we expect to find?

Suppose that hard M-reflector is indeed the buried surface of an old desert that once stretched the entire length and width of the Mediterranean Sea. It would be a fearsome place: hot desert winds, whirling and tossing dusts and salt across an enormous desert basin, its floor more than a mile below sea level. It is nearly the size of the United States, as though many hundreds of Death Valleys or Dead Seas had formed side by side to endure for a considerable time. It was no single monotonous landscape, but a whole varied region, crossed and marked by volcanic mountain ranges, their low foothills sloping to flat valleys and deeper basins. Here and there salty marshes and lakes, large and small, would form and dry again as rainfall and rivers from the continental uplands flow and change. Seawater from the ocean high above to the west might leak in. All those bodies of water would have shifting shorelines, beaches, flats, now dry and again awash in long sequences over time. The right geologists could recognize all those processes; indeed they would see more, because they bring to the study the result of detailed examination of the salt desert shores of today the world around.

Within a few days after they first read the astonishing gravels from the M-layer our enthusiasts had a working theory. Their ship was no longer in crisis, seeking vainly for uncertain clues; it was a powerful instrument of science in their young hands. They led it to site after chosen site at sea for six weeks, to retrieve core after core from below the M-layer. They could test the detailed predictions that they knew their picture of that great salt desert implied.

A drilled core is not merely a loose sample of rocks, like a handful of gravel. It is a cylinder cut out of time, at one single spot where there took place one single sequence of events out of a long past. In the core the context is in place; it surrounds every feature the geologist can recognize: what came first, what next, and what next again, as you move upwards along the core. The evidence is there before your eyes; to be sure, each stretch of textured rock has to be given its meaning out of the specialist's rich experience with modern analogues of the ancient material you see.

Follow the passage of time along one of the first good cores taken out of M-reflector. The vivid account of the past is supported by direct evidence at every point.

The drill that here probed well below the

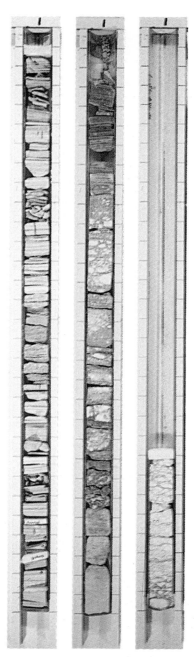

hard M-layer stopped as it entered into rock that was now a dark fine carbonate mud that had slowly accumulated in the tranquil bottom of a deep lake. Suddenly that lake dried up, and a little higher in the core we are in the thick salt flats that had formed on the dried lake bed. Next in the core is an unusual light rock, patterned with nodules of white material separated by thin wisps of gray. The Texas oilmen had long ago named it "chicken-wire anhydrite." Those white nodules are the mineral anhydrite, another very dry form of gypsum, drier even than the powdery plaster of Paris. The characteristic markings that gave the structure its homely name are well-known to the petroleum geologists—Ken Hsü among them—who have walked and dug today on the hot arid flats on the shores of the Persian Gulf. The recognizable structure slowly forms in the ground from the evaporation of salt water under extreme drying conditions, on a flat shore that is only occasionally awash with water, with virtually no rain, and air temperatures above 100 degrees. That combination of conditions almost uniquely makes this unmistakable but unfamiliar form of a familiar substance dissolved in all seawater.

Go a few feet more up into the core. The lake recorded here does not stay permanently dry. It floods and dries over and over again in cycles. Here it lays down gray thin horizontal layers, fine silty muds, and then it is again des-

Three cores from below the M-reflector, each five feet long. The core materials and what they record of the history and life of this ancient salt lake were read directly from the cores by Bill Ryan. The leftmost core is the deepest and the oldest. Between the middle core and the last one there is a gap of about fifteen feet.

An ocean in a bowl dries up indoors, seen in several stages as the water evaporated and the glistening salts came down.

iccated and exposed to the hot air. Those strange anhydrite nodules grow once more in the surface deposits where once was salty lake water.

The shallow basin floods yet again. Storms, winds, tides push the water across its margins, laying down a stretch of finely laminated rock. If we remove a bit of the lake deposits from the core, to scrape the surface gently with a razor blade, we expose dozens of elegant paper-thin laminations. These record a yearly rhythm on the now rather deep lake bottom. Their darker portions are formed when the spring floods carry in much water with a load of coarser particles. Those alternate with lighter, finer deposits that occur during the quieter times of fall and winter. The uniformity of all these layers, like so many book pages, records that the lake was lifeless during all this time, no worms or snails, no crabs or crawling creatures to stir up the mud. It was as lifeless as the Dead Sea, and for the same reason: too much salt in the briny waters of this desert lake.

Next we come to crinkled surfaces, wavy surfaces in the layers. Here we see for a while a record of shallower waters, where the winds were able to stir up currents to winnow the shallower parts of the lake floor. Then the lake grew deeper and quieter again, and shallower once again, with more ripples. Now the surface was newly dry and exposed. Signs of waterborne gravel and wind-drifted dusts and sands are here.

The Rosetta stone for this entire environment is in the core, too, a few tens of feet above the lowest portion. It is an unmistakable formation of layers to be found today on many half-dry, half-wet flats along warm coasts. The formation is called stromatolite; it is in a way a fossil. Within this layer are the remains of blue-green algae. They required sunlight for photosynthesis. The algae lay down a mat as they grow, a living mucus surface. Gentle waves send a little water across the algal mat. The sticky mat traps particles, and the deposits build season by season. Here was absolute confirmation that the deep abyssal floor of the Mediterranean had seen sunlight and been exposed, to be flooded again, to dry and flood again. The core ends with cracks in the solid deposits of salty muds; the salty surface has been sun-dried for a considerable time, shrunk, and finally cracked: a salt desert floor.

We could follow many cores through hundreds of such small local cycles of getting wet and drying, bunched perhaps into a dozen major cycles that occur throughout the whole Mediterranean.

Such a core had plainly probed the margin of some desert basin during the history of the dry sea. But if those flats were sometimes dry, sometimes wet, then the higher mountain slopes around them must have been well above water. The drill was sent down later in the voyage to probe just those slopes. It cored through hundreds of feet of land deposits from flowing streams, essentially soils, gravels, clays.

The greens and the reds, the colorful dusts that paint the desert, are there, too. Ken Hsü was excited when he saw in the cores the very substances, the very colors he knew in the desert foothills above the salt flats of Death Valley. There long-vanished streams had brought down gravels and sands, as the floodwaters fanned out to evaporate on the desert floor. In other places on the slopes there lie windblown and water-borne silty soils. The core was a presage of modern Death Valley. If Death Valley is so well imaged on the floor of a sea, that deep sea must have spent a long time close to bone-dry.

CORES ARE GRAPHIC, BUT THEIR DETAILED testimony is local, restricted in scope. The geologists knew that they would have to address the evaporation of the Mediterranean overall.

Once a whole ocean of seawater is dried up, something is left behind. I tried out for myself what would be left from half a bucketful of Boston Harbor water; all the oceans are quite well mixed. It took a couple of weeks to evaporate indoors from a wide bowl I filled and left on the table in the spare room. The residue was a thin glittering crust of white crystals, little cubes and square plates, many so small it was hard to make out their form. There was certainly plenty to interest anyone able to recognize the distinct minerals in the deposit. Most of it is plain table salt.

There is one useful overall conclusion from

that simple trial. Although the processes at tabletop scale are not just the same as on the shore, the weight balance is reliable. It takes plenty of seawater to yield much salt upon evaporation. It turns out that a foot layer of salt requires evaporation of fifty or sixty feet of sea water.

The drillship had taken cores repeatedly through the M-layer, into the flat margins of a basin and into slopes of undersea mountains standing well above the basins. The co-chiefs and their friends would certainly not neglect to seek out the deepest part of some old salt lake. They could expect that water would have filled the deeper basins that they could reach, the last places to go dry in the changing fortunes of an evaporating sea.

The last drill site of the voyage was placed on the margin of the abyss, not far from the deep flat bottom of the basin west of the island mountains of Sardinia. There as they finished the cruise at least they had nicked the bullseye! As evaporation of seawater proceeds, what is left behind is a more and more concentrated brine. A known and regular series of salts deposits out of the solution, the not-very-soluble gypsum among the first. Toward the very end the now concentrated brine lays down the most abundant of seawater minerals, sodium chlorine or common table salt. Last of all to come out of solution, from residual bitter waters rich in the most soluble of the salts, is potash, the potassium chloride found in every garden fertilizer.

The drillers brought up a solid core of rock salt, just as they had hoped, from almost two miles down. They had cored not far away from the center of one bullseye, the legacy of an ever-diminishing saline lake that became smaller and smaller as evaporation went on. But during the return cruise into the Mediterranean, made in the spring of 1975, the drill sampled the center of the deepest basin of all in the old floor of the desert, to core a big lens of potash, perhaps the largest such deposit known, left behind from a great salt lake that had dried there below what is now the Ionian Sea.

At the end of the 1970 cruise the happy geologists went back home to work. But the *New York Times* had reported something of their spectacular findings. Bill Ryan soon had a letter from a Soviet geologist, I. S. Chumakov. Years before, Chumakov related, he had been part of a team whose tasks included drilling deep boreholes to test the rock under the site of the great Aswan Dam that Soviet engineers would soon build across the Nile. The drill holes penetrated the river water, then the freshwater sediments, down into marine sediments hundreds of feet below present-day sea level, and finally through the later sediments that had filled an old narrow gorge cut by erosion deep into granite bedrock, its bottom fully

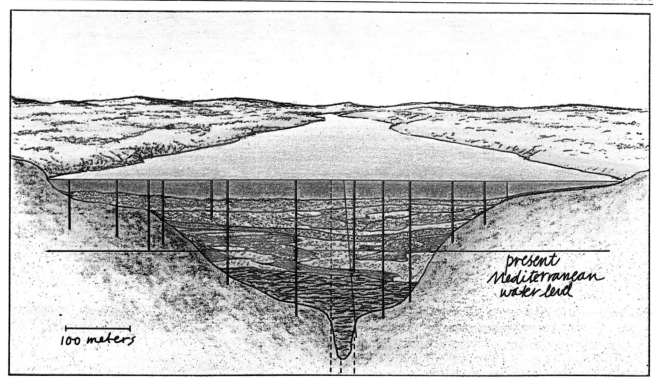

100 meters

present
Mediterranean
water level

600 feet below present sea level. The interpretation was clear: the sea must have been much lower at the time of that erosion, so the Nile could flow swiftly north to the remote desert basin, cutting a deeper and deeper gorge as it went well out into what is now sea. Below Cairo near the present seacoast a similar buried narrow gorge has been found recently during exploration for oil. It is now filled with later sediments, but it was once cut a mile and a half deep below the present level of the sea!

The Rhone River too has a hidden canyon, and there are similar deep incisions on land and across the slope of the continents for rivers all around the Mediterranean. The desert had left its mark far inland. With that mark we gain some indication of the time it had endured before the sea returned to fill the desert basin, and to catch and tame the wild rivers at an elevation far above the depths of their buried canyons.

The preposterous desertification has found plenty of detailed confirmation, core after core, and inland, too. It is time to question more widely; there are balances that must be struck. It was the water balance that first

Under the high Nile Dam at Aswan the Soviet geologists had probed the deep rocks with fifteen drill holes, marked here as vertical lines. Their map of the subsurface is redrawn here true to scale, a cross section of the river valley as it is today. Notice how high the Nile surface stands above the level of the sea, now 800 miles downstream. When the old Nile cut that lowest furrow deep into the granite below, the river must have been running downhill much farther to sea level at that time, a level far below the present. The Med had dried up.

assured us that the Med could dry up, that such an event was not precluded.

Now for the salt. It was long well understood that the region is a salty place. The seismic profiles had picked out thick layers and domes of salt below the seafloor, and there are classical salt mines on land, for example on uplifted Sicily, with thick visible beds of layered rock salt. If you dry up one entire seaful of Mediterranean by closing the straits, you would not reclaim anything close to the amount of salt you find below the floor today. We know the volume of salt present. It would spread into a uniform layer of sea salt about 800 feet thick. But the depth of the seawater over the same area averages under a mile. My bowl of Boston Harbor water evaporating on the tabletop measures the need: seawater enough to carry in the salt we find would fill the whole Mediterranean ten or a dozen times over.

Could the performance simply have been repeated *da capo* as the valve at Gibraltar opened and closed a dozen times? Conceivable, but bizarre. There is a sensible, even obvious, alternative, already pointed to by the shipboard paleontologists. In many of the desert cores the organisms present are the little animals of shallow coastal waters, of lagoons, even of inland salt lakes. Those cyclical and shifting desert lakes were not deep; instead, they were always rather shallow sheets of water along the floor of the mile-deep basin. All that salt had accumulated slowly, as a series of slow leaks of ocean water from the west had fed water to the floor of the basin, to dry there for a long time. The Mediterranean acted as a giant set of salt pans, lake beds large and small, where now one and then another shallow body of water evaporated to fill some local depression with salt, to be replenished slowly again and again. There is time in plenty while the wide M-layer desert endured.

The old blue sea had dried up into a white salt desert. The beginning of the change might have been gradual, the Straits restricted, the sea growing more and more stagnant. But once the great currents to and from the Atlantic had substantially stopped, the entire sea dried fast, within a few thousand or tens of thousands of years, very speedily indeed compared to the millions of years that have passed since that time when the M-layer was formed. Act One of this drama was brief. The sea was soon dry.

Act Two was a long, slow one. The lake-studded desert was fed by a slow inflow until enough water had been brought in to supply all the salt we find. That may have taken a hundred times as long as the initial drying; the inflow had to remain limited, at a modest average rate so that evaporation could keep pace. We do not know just where those inward leaks occurred. Near the new Isthmus of Gibraltar seems probable, and quite recently hints of long-lasting small inflow may have been found across southern Spain by Ken Hsü and his students.

Certainly the whole deep basin is no desert today. What happened to fill the sea up again? Act Three came as an abrupt dramatic climax.

One of the 1970 cores was brought from the deep seafloor between Sardinia and Italy. Maria Cita recalls the very core: "It was a perfect recovery. For the first time within a single core we found a contact between two contrasting sediments. There was a boundary sharp enough to put your finger on the contact. The environments were entirely different, but they were adjacent. There was nothing missing in terms of time."

Frederick Edwin Church: *Niagara* In the collection of The Corcoran Gallery of Art, Museum Purchase

Below that contact were the familiar evaporated salts and muds, the residue of the salt desert. The organisms there would be dwarfed, specialized, the creatures of shallow and brackish waters. Above the contact were the reddish-pink colors of a deep ocean ooze, well-oxygenated and cold, full of a normally diverse collection of oceanic organisms.

Finally there must have come an earthshaking movement. The huge dam at Gibraltar broke, and the waters rushed in suddenly, to refill the basin with Atlantic waters and their organisms. The rate of inflow must have been great enough so that the evaporation was overwhelmed and the seas could return, with no significant drying to delay the refilling. We know more than that; everywhere in the basin the first organisms that appear are already those of cold deep Atlantic waters. The lip of the Falls of Gibraltar must have been well under the level of the Atlantic, maybe as much as a thousand feet. Under that enormous head of pressure the waters roared and cascaded down, perhaps a thousand feet and more to the first terrace, then over long cataracts down into the empty mile-deep basin.

What a thundering roar! A hundred miles out into the Mediterranean from Gibraltar the seafloor cores show a gap in the record. There is no sign left there of the M-layer, evaporation, the salt deposits, of the whole period of the desert basin. A few hundred feet are missing,

Niagara made a splendid sight to the American painter Frederick Edwin Church, who reported the Falls this way in 1857. The Falls accept the drainage of the Great Lakes, keeping up rather well with what water comes into the lakes by rainfall, snow, and rivers. But this torrent could never fill the Mediterranean Basin, for it falls far short of meeting the net loss of water there by evaporation.

from the M-layer and below. That thickness of seafloor must have been washed away, scoured clean over a vast area by the terrible force of the thunderous flow. The apes, our cousins, ought to have been startled by the roar, but who has gathered their stories? No hominids we know of were yet present.

There is an estimate that the incredible falls could not have lasted for much more than one human lifetime. The isolated Mediterranean waters, without the Atlantic connection, would stratify in temperature, like an unmixed pool, the bottom waters cold, the upper layers sunwarmed. There are means for measuring the temperature of the waters in which some organism has grown a shell. (The effect is a subtle one on the chemical reactions of growth; the temperature affects the relative rate of incorporation of two oxygen isotopes in the material of the shell.) The very first millimeters of a core that records the newly returned Atlantic fauna show no difference in growth temperatures from those higher up. Nowadays the ocean stirring mixes the sea, and there is only a small temperature difference. But if the return had been slow, the waters would have had time to stratify, to show signs of a cold bottom layer and a warmer surface. That they do not means that mixing was present from the start. It would take about a hundred years to set up the stratified sea. So the currents were not delayed more than a hundred years or so after the water began its return. That is a geological instant. If we consider all the time since the gate first closed one whole day, then, all in proportion, the sea first dried out in about a minute, the seafloor desert and its shifting lakes lay under the sun for an hour, until the dam finally broke, and the prodigious falls appeared and ended all within a second or two. For the rest of that long day until now, the blue Mediterranean has been a quiet pool of the world ocean.

On top of the ridge of the Rock of Gibraltar, Bill Ryan and I looked down to the flat sea and reflected on the ancient falls. We spoke of many points. One was the possible response of the world ocean to the storage of all that salt in the desert Mediterranean basin; did that presage the coming of the glaciers, as the polar waters lost a little of their antifreeze?

I asked if the Falls of Gibraltar were really unique; was there only once so swift and surprising an event? He told me of a couple of possible examples elsewhere of just the same drama. One was very long ago, when the Atlantic itself was a constricted narrow sea, still forming. An older ocean sheet, called the Tethys Ocean, then supplied water to the growing Atlantic from the east. Was that cut off once? The telltale salt beds are there, split by the later growth of the Atlantic into two halves, one deep undersea off Morocco, the other off Nova Scotia.

Ryan had never been on the Rock before, though he had spent many an anxious hour not far off at sea, plotting its lighthouse on his maps to help locate his ship. His first visit to the Rock moved him to question. The Rock of Gibraltar is limestone, product of shallow seas. No one finds it strange any longer that what was low can be uplifted. People, geologists and laymen alike, are used to seeing fossil shells on mountaintops. But they do not easily accept the notion that a whole deep ocean filled with blue water might once have been desert dry.

Yet water is the most mobile of abundant minerals. The glacial ice that once covered northern Europe was water; it came and it went. Why not the seawater we see here around us? It is about the same in amount as was the ice over Europe. Yet "the well-educated geologist finds it incredible."

"They'll come around in time," I said.

Most of the early reluctance to accept the story—by now it is very widely accepted—came from the suddenness of the main events. Geologists are used to long, slow processes, and they are puzzled and worried by speed, for they spent a century fighting off the prejudices of an unscientific chronology that had reckoned the earth at only 6,000 years old, off by a millionfold. A physicist, perhaps more detached, sees that the world displays slow processes and swift ones, even together in one

The Great Falls of Gibraltar drawn to scale at one estimate of their size. No such falls of salt water are known today. The small rectangle marks out Niagara to scale. The largest of present-day waterfalls, Victoria Falls in Uganda, spills ten times the volume of Niagara, still minor alongside Gibraltar.

single context. The rocks of the canyon wall erode, grain by grain, and then once in a while a whole boulder suddenly falls. Both processes are important. There is little to be said for holding a preconception about rates in a world that clearly has many scales of size, and surely many scales of time to go with them.

THE PROCESSES OF EARTH HAVE NOT BECOME static just because we are here. We know the continents still slowly drift and move. We know that the sun still dries up seawater, leaving salt behind. It is only that the time scale on which continents shift is long compared to our lives, even to the whole of human history. But if human beings survive long enough, it is quite likely that it will all happen again. The valve at still-narrow Gibraltar will close, the entire blue Mediterranean will dry up until a few shallow salt lakes will occupy the deep basin of the sea. Then the braver tourists will want to travel by camel train from Cairo across the hot salt desert to the high mountains of Sicily!

That is a narrative that the new knowledge and the new instruments of the geologists have brought us. We have had a chance to read a few pages out of the long history of earth. They dis-

close a past that, different as it may be from the present, is always somehow recognizable.

The way we examined that past has changed sharply with the growth of science and technology, with more powerful methods to search for clues in the world, to measure time, to examine the nature of rock. But the chief idea remains the same: seek more evidence in the world and reflect as widely and carefully as possible upon what you find. That method is the same in the wide valley as it is beneath the seas, with volcano-stricken forests as with the birth and death of oceans, or when someday the geologists will probe deeply the surfaces of Mars and the moon.

It is the same exciting way of working that a hundred years ago allowed a few astonished geologists, carrying little but hammer and sketchpad, to walk the wild lands, to note trees of stone that stood one above the other, and to puzzle out how that marvel could come to be.

5

ATOMS

THE ROCKET'S RED GLARE, LIKE ALL THE brilliant colors that part glorious fireworks from commonplace fire, is the direct consequence of specific substances included in the glowing material. It was the eighteenth century that first saw such dazzling firecraft, as the chemists began to recognize the diversity of matter. That tight link between the color of certain glowing matter and its chemistry has brought us not only the recipe of the distant stars, but rather surprisingly handed us a master key to the architecture of the atom itself.

Within a small brass tube I have you can look at any time to see an unceasing play of tiny sparks, a more intimate sort of fireworks. They are in fact atomic fireworks, more spectacular to the mind's eye than any blazing skyful of color. These fireworks are not spread across the sky, but only over a stamp-sized screen painted with a mineral pigment, not unlike the screen of a television set in miniature. They are faint; a watcher can see them only in pitch darkness, eyes accommodated to the dark. So faint are they that the film camera can record them only with special aids: a modern night-vision amplifying device and a high-sensitivity video camera. But the little flashes lie just within the margins of dark-adapted human vision.

They are marvels, because they directly connect human perception to individual atomic events. That sudden, random, particulate sequence of sparks flashing all over the screen is the sign of the atomic nature of matter. It was the first visible sign that we ever had.

The device we can look into in the dark was the remarkably simple invention of an ingenious physicist in London around the turn of the century. The one shown here is marked with the year and signed with its maker's name. Sir William Crookes had named his contrivance the spinthariscope, from the Greek word for spark. This example was sold as a kind of scientific toy; it was a wonder in that exciting time when radium was new.

The brass tube holds an eyepiece, a simple magnifying lens that enlarges five or ten times. That lens is focused on the fluorescent mineral screen that closes the other end of the tube, an inch or two away. Mounted just above the screen is a small needle. On the needle's end there was placed back in 1903 a tiny sample of a radium compound, obtained by dipping that needle into a very dilute solution and allowing the point to dry. In those days radium might

Fourth of July fireworks.

have sold for a million pounds sterling per ounce; this sample could not have been worth as much as sixpence.

That sample on the needle tip is unweighably light and invisibly small. And yet from it come every one of those sparks, roughly a hundred a second allowing for what the eye misses. They have been coming for close to a hundred years without noticeable change. That means about a trillion flashes since this device was new. From the work of the Curies it was recognized that this process would continue for a couple of thousand years. It was known as well that radium thus slowly changed to an inert element; each visible flash—we don't see them all, and not all events make a flash—must have been witness to the decay of at least one atom of the sample. That may give some feeling for the innumerable atoms that crowd all matter.

That the element radium glowed in the dark had been noticed in the first pure samples the Curies made in their cold Paris shed. That glow was most luminous when the particles of matter flung out by radioactive radium were allowed to impinge on the surface of any of a number of known fluorescent minerals, like the zinc sulfide used for this old screen. Sir William knew all that; what he did for the first time was to examine closely with a low-power microscope the fluorescent glow that came from a minute sample of radium. What he saw

was the startling sight he shared with us through his spinthariscope: that faint luminosity was not at all a steady continuous glow, but could be clearly made out as a restless succession of individual sparks.

The flashes we actually see in the spinthariscope are not each the light from a single atom. The comparison with fireworks is helpful here. A single rocket initiates one complex glowing path across the sky; here a single very fast moving atom, the decay product of radium, plows its way through many tiny crystals of the zinc sulfide screen, to excite light from a great many of the atoms it passes. The initial event is single, like the rocket, but enough light for the eye to see requires the contribution of light by a great many atoms, each of which was in its turn disturbed by the swift passage of the one rocketlike atom. These are not commonplace atomic events; radioactivity is no ordinary atomic process, but one that is enormously energy-rich. That is why the eye can follow a single radioactive decay.

A host of atoms must be present in this minute sample. Here their number is disclosed by the special nature of the substance involved. Atoms are remarkably numerous, and therefore tiny, for so they must be if trillions crowd onto the point of a needle. The events in which they take part are sudden, random, and discrete, the very nature of the atomic world.

The spinthariscope first revealed atoms

The brass spinthariscope of 1903 holds the faint sparks. Those sparks could be counted one by one, and the little demonstration device was in fact soon made into a quantitative instrument for the study of radioactive decay. It was by laborious counts made in the dark of the flashes in such a scintillation detector, the finest instrument of that pre-electronics era, that Ernest Rutherford and his colleagues in 1911 discovered the nucleus of the atom.

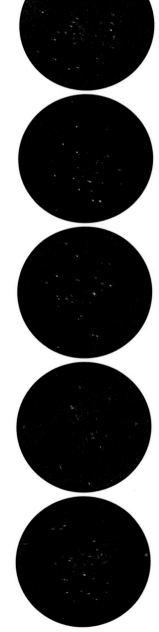

directly to the eye. But its flashes cannot of themselves be called convincing. You must be well prepared to interpret those minute fireworks, as a few chemists and physicists were at the turn of the century. There were doubters then even among learned scientists, though the directness and simplicity of the evidence from the spinthariscope shook many skeptics.

We propose to spend the rest of this chapter walking diverse paths toward understanding of the atom, for no single route is in itself a proof. Some of those paths have been walked for thousands of years by the craftsmen; some of them are newly found in the laboratories of our day. I hope to show that all together they lead to a single central point, the compelling conclusion that all matter, even your own hand, is at sufficiently small scale genuinely grainy and atomic.

Purity

URBAN BILLMEIER OF CHICAGO HAS SPENT A lifetime as a master craftsman. He works with pure gold in a craft tradition that can be traced back 5,000 years. He is a gold-beater, one of the last in the United States to follow the classical method by hand.

BILLMEIER: We start out with 24-karat gold, pure gold, and add enough copper and silver to give it just enough hardness

Atomic fireworks, incessant, faint, random flashes or sparks within a little tube, as photographed by an electronic imaging device. We can estimate the number of the flashes, even though in this example they come too fast to count well. A few trillion potential flashes were put in this tube back in 1903.

that we can handle it. But it will still test at 23-karat plus.

The people we sell to are in a demanding field—let's call it permanence. When they re-finish a very expensive antique, they want that to last another hundred years. If it is a picture frame, again, a hundred years. Maybe it is a gilded dome on some public building; they might not want to do that over again for fifty years. The gold they use has to have the right wearing qualities. It cannot discolor, and it must have enough body to stand up under the elements.

It's just the properties of gold. Gold is a fabulous metal. There is just no end to the uses of real gold.

Gold-beating by hand is a long and highly skilled procedure for turning pure gold metal into the extremely thin yet near-flawless sheet called gold leaf. There is machine-made gold leaf as well, but if it is beaten really thin it contains more flaws, for while they hammer, the machines cannot correct as the craftsman can for small imperfections left from preceding steps.

A small ingot of gold, very slightly alloyed, is rolled out between steel rolls, over and over again, into a long thin gold foil. A witness need not even watch the process but only listen, to hear the metal grow thinner and thinner as it falls and folds from the rolls. At first it thuds into the bin, as dully heavy as cardboard, then it sounds crisp as paper, until at last it is tinkling foil. No other cold metal can be thinned so much without tearing and cracking; this is gold.

The rolling mill is not as old as the gold-beating craft itself, though its use for metals can be documented through nearly 500 years. Rolling lengthens and widens the metal; it thins in proportion, the volume of gold staying about the same. The rolled strip ends up as fifty feet of inch-wide thin gold ribbon.

But that is only the start. Now the foil is cut into small pieces, and each piece is sandwiched within the parchment leaves of a sort of square book. The book is beaten for twenty minutes until the pieces are each about the size of the small page. Each piece is then cut into four, and the process repeated, now by beating for a couple of hours.

Once those sheets have spread in turn under the beating until the pieces fill the pages of the second book, the gold is removed, and again each sheet is cut into four pieces. Finally the feathery gold is placed between the leaves of a book whose pages are made of goldbeater's skin, a particular portion of the intestinal lining of cattle. That membrane is extremely resistant to pounding, and yet gives so as not to tear the delicate metal; here the echo of antiquity is strong and clear. The beating continues for four hours more, the book turned systemati-

SCENES FROM
THE GOLD-BEATER'S ART

The rolling mill.

Through the mill.

Gold ribbon.

Cutting gold foil.

Filling a book for beating.

One stage of hammering a book.

Feather-thin in a book.

Removing the thinnest leaf.

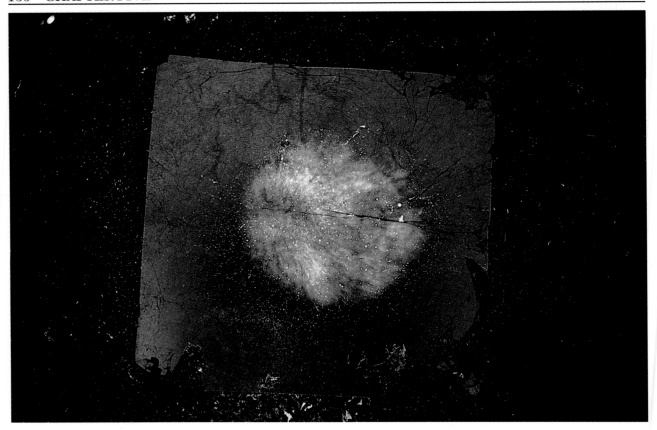

Gold leaf. The glitter of gold in a metal sheet so thin that it is translucent, passing on as blue-green a white light from behind.

cally as it receives measured blows by hand from the large wooden mallet.

The final book opens to disclose the finished leaf, too fragile to handle at all except with breath and soft brush. The metal is still perfectly golden in reflection, though a translucent blue-green in strong light from behind. The final leaf is so thin that were there a complete Bible made of gold-leaf pages it would turn out no thicker than one single leaf of ordinary Bible paper.

There are many specific and elaborated technologies of worked substances like clay, dyes, metal, glass, leather to be found in the workshops of the handcrafts. Gold-beating is so old and so cosmopolitan, so magical in effect, that it merits special attention. But all work with gold has been remarkably mature and powerful for a long time. The artisans every-

where came to understand that the noble material itself was the basis of their skills, that it would respond reliably to careful treatment, that the gold from any source, if it were suitably pure, could be worked in the same way and would endure. Even a part in fifty of alloy made a noticeable difference, if sometimes a difference to be prized. No metal but nearly pure gold can be so arduously worked and thinned without hardening and then cracking.

GOLD WORKERS EARLY CAME TO REALIZE that their precious material remained recoverable even after processes that very much changed it visibly. Gold dust could be hammered or melted into some desired form of the solid metal; it was important to know how much had been lost or added. The first use of fine weights and balances is associated with gold-working as far back as pre-dynastic Egypt, more than 6,000 years ago.

Craftsmen also learned how to purify impure gold for their use. The content of the

An assay laboratory in the time of Faust. The master of this quite unmagical lab of about 1570 was equipped to assay and refine copper, lead, silver, and gold, to recycle his mercury, and to make needed acid. Cupellation was carried out in the assay furnace in the room to the right. The author of the old book from which this print comes points out that the assayer should be well versed in arithmetic; "indeed, it is the very sign of a master."

essential stuff could be judged through control of the chemical process and by careful weighing. Such assaying of gold is very old. The fire assay method used in Mr. Billmeier's shop, called cupellation, has been in wide use for 3,000 years. Until a decade or so ago it was still standard procedure.

The assay begins by adding a surplus of granules of metallic lead to the test sample held in a special little porous cup, called a cupel, which will contain the melted mixture in the furnace. The cupel is molded out of the ash of wood and bone. Once the lead is red-hot, it reacts with the air. The molten fluid that results, an oxide of lead, is able to dissolve most of the similar oxides that form at high temperatures from lead and other metallic impurities in the sample under test. That lead-rich fluid, the dross, but not the gold, also molten at this time, is then blotted up by the porous ash; a used cupel

is visibly stained by the dross it has absorbed. The little bead of gold (with any other noble metal that may be in it) remains behind, unaffected by the air, to cool, solidify, and be carefully weighed.

In such ingenious crafts lies a great part of the ancestry of the sciences of substance, a legacy rich with pointers for the chemists who were to come, even to hints of the unseen atom.

Molds for making ash cupels of the 1570s, and cupels drying.

A master of today weighs a test sample for fire assay by cupellation.

The molten sample.

The white cupel is stained by dross, but it has retained all the gold as a purified bead, to be weighed again.

Sorting Sand

A GOOD BEACH IS A WIDE AND GENEROUS place, extending from the sea's edge to the green fields. It is large compared to any person who roams it. Were it small, it would not work like a beach. It would not respond as expected to wind and water. That breadth is important. What we see at the beach would not be there in a tiny space; a pinch of sand is no beach.

Yet equally a beach is not a rocky ledge. It is mobile; we often see it moving under wind and water. There is no beach without these two scales: the sizable open expanse, and the myriad, repeated small grains of sand that inwardly confer motion.

To look more intimately at sand as a substance, we may seek the aid of a magnifier. One look is persuasive. Sand is one word, but the beach sand is not one stuff. Under the lens one can see black grains, pink grains, and clear, whitish grains. Here and there you can notice even at a distance that waves and wind have somewhat sorted the diverse grains of sand by motion on a larger scale than in my hand.

We are creatures of the eye; it seems natural to us to sort sand grains by size and color. Plainly the method works, but certainly there are many other ways of sorting.

One rather surprising way is with a small magnet. It sorts grains because some of them love to adhere to it; from this particular sand the magnet tends to collect black grains. We have no reason to believe that the black grains

Natural sorts.

Sand sorted naturally by wind and water at the beach:

paler sand, darker sand, and sand gleaming with little plates of mica, seen also through a hand magnifier.

The magnet picked out black grains of beach sand.

we sorted by color are the very same ones that the magnet picks up; the magnet may take only a fraction of all the black ones. The point is easily made: our sorting is never perfected, never complete. We must try property after property until we are satisfied that a single sort would imply a single set of properties that all the grains would have in common. Then we might speak of pure sand of that grain composition. We could never be sure that some other untried process might yet part the grains that had so far seemed identical. But this explanation, that the purity of a sample of sand is a tentative singleness among very similar grains, is a strong hint from the beach about the inner nature of matter.

The world beyond the beach behaves that way, too. Pure water has one flavor. Pure gold is admired by craftsmen everywhere because it works so well in one single way. The sand grains surely are not atoms; but the idea of sorting to what we judge as a single set of properties resembles the world so much that it suggests that within the world of matter, as indeed within every grain of sand, there is another invisible graininess, the graininess of the atom.

It is easy to divide beach sand into distinct, purified components. When we did it by hand and eye, or by a magnet, we recognized the operation as a sorting, grain by grain. But when the assayer purified the gold sample in the little cupel, there was no trace of graininess anywhere. The gold bead had a smooth and lustrous texture. And yet might it not be that at a scale too small for us to see the two kinds of purification are really one? Somehow within metal there may be grains—call them atoms—that have been sorted under the processes of the assayer as the natural forces of wind or magnet sort the sands of the beach.

Elements

WHEN I WANTED TO TRY ANOTHER CHEMICAL sorting even older and more familiar than the assay of gold, I sought the help of a generous neighbor, one who is a celebrated applied biochemist.

"I'm Julia Child.

"I'm doing a very special meal, cooking it just the way Phil has requested. We have a very colorful meal here, I think, with these beautiful fresh green beans, this bright orange squash, scalloped potatoes, a tenderloin steak with a little mushroom sauce on top. I'll just finish that off with a little more green for color. Here's an extra Phil touch—I'm going to do it just the way he likes it—and that is with a teaspoon of coarse salt right in the middle of the plate. Then the final Morrison touch, preparing it in a particular way in the oven.

"Now I'm going to take the main course

Carbon. Foodstuffs are ores of carbon; the white salt is an exception.

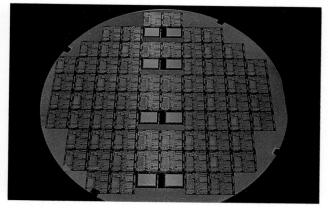

Silicon. The wafer on which the intricate circuits called computer chips have been formed is a crystal of pure silicon, the purest element that has ever entered into large-scale industrial production.

out of the oven. Ah ha! That turned out very well, too. Nothing like being an experienced chef in these matters. Look at that: a meal fit for a king. There you are, Professor Morrison—this is Julia Child. *Bon appétit!*"

Julia Child had generously produced with characteristic good humor, skill, and aplomb a meal made absolutely to order, every dish of which had charred to a crusty black, not by any inadvertence but by design, to make a point about substance.

Such an atomic sorting is as commonplace as the morning's burned toast. Julia had been able to produce a similar black layer over the surface of half a dozen distinct foodstuffs. To be sure, it didn't take all her artistry to burn foods. We can all do that. But the common consequence holds something of note. It demonstrates that whatever we eat is an ore of carbon. Salt is an exception, and water, too, but they both play specialized if essential roles in nutrition.

The skeptical chemist would want a bit of checking before she would accept that these black crusts that all look the same are in fact the same substance. She might take dried bits, one sample from each dish, to burn again more completely in oxygen, to check chemically that the products of complete combustion were all one. They would be closely alike; the chemist

would conclude that carbon—a word that is only the Latin name for the common black fuel—is indeed found in nearly all foodstuffs. That familiar yet sweeping result, that a single component can be extracted from a great many related source substances, is one of the foundations of our understanding of the atom.

A CLASS OF SUCH SORTED GRAINS IS WHAT WE call a chemical element. We can ourselves break some well-known material into its elementary constituents. With some rather strange-looking apparatus of glass tubing I easily took apart and put back together again the substance water.

Water, too, is easily sorted, sorted into two well-known elementary gases by the passage of electrical current. The direct current we used entered the water near the base of one side tube, to leave by the other. (A small amount of acid greatly increases the current the fluid will pass.) In both tubes bubbles formed to rise through the water column that filled the tube, bubbles of hydrogen in the left tube, bubbles of oxygen on the right tube, a result that would be easy to check. The copious little bubbles summed up to form a big gassy space at the top of each tube.

The proportions of the two gas-filled spaces stayed constant as the bubbles formed steadily and the water levels dropped. The most famous

Water analyzed, 2 H for 1 O.

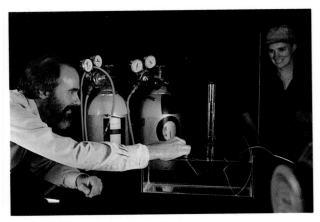

Water synthesized, 2 H for 1 O.

of chemical formulas was being realized before our eyes: two volumes of hydrogen to one volume of oxygen—H_2O. Something internal must be behind that simplicity. There is no means by which the apparatus could build in a simple ratio.

Less often seen is the exact reverse of that process, now combining the two gaseous elements together to form water. Two steel cylinders each held commercially pure gas under pressure, the one hydrogen gas, the other oxygen. We placed a mixture of equal volumes of the two gases into a single thick-walled glass tube filled all the way with water, its mouth inverted under the water surface. First we added one gas and then the other, bubbling in turn from each of the two source cylinders. We tried to admit quite equal volumes of each gas, which in the end jointly displaced a little of the water to open a small gas space at the top of the tube. The two gases combined to form water whenever we triggered the reaction in the top of the tube by an electrical spark that ignited a little gas explosion.

The small amount of newly made water simply joined the water already there. (The explosion was interesting to watch in slow motion; the film verified that none of the gas leaked out of the submerged mouth of the tube even during sudden explosive motion; it all stayed confined.)

Hidden jewels of quartz lay within this geode; what will be found in the next one?

But there was some gas left over in the top of the tube. There was *always* some gas left over. We tried all one afternoon to vary that result, but the gases were stubborn.

We had marked the tube top to indicate four little thimblefuls of gas, four equal volumes. Before ignition that space had been filled with equal volumes of hydrogen and oxygen; two thimblefuls of each gas had been placed there. They always reacted vigorously to leave behind a residue of exactly one thimbleful: for two volumes of hydrogen will take up only one volume of oxygen to form water. The residue was simply the amount of oxygen that could not be taken up into water, always one volume of the total of four. Again we learned H_2O. Something deep within water appears to know simple arithmetic.

Whether we are taking water apart or putting it together, we find that the process insists on the same result: in water two volumes of hydrogen accompany each single volume of oxygen. I could not claim there is only one way to explain so simple a result, but if there are atoms, and if within water they do cling into identical little clusters of the right type, two of hydrogen always bound tightly to one of oxygen, that would explain it all neatly: H_2O. (That minimal bound cluster of atoms might also be H_4O_2, or H_6O_3, or any other similarly related value; we cannot distinguish among those on the basis of these two experiments alone.)

Form

UNTIL THE TWENTIETH CENTURY THE CHEMists pursued their way to the atom they could not see by rather subtle means, independent of atomic visualization: their glass apparatus that held so many reactions, the sensitive balance, and the arithmetic of small whole numbers that could be puzzled out of all those weights and volumes. But there were other clues in matter that the eye could follow, once they were recognized.

A small natural spheroid of rock like the ones in the picture is a striking but by no means rare mineral specimen; they are found in many places. I could crack one open at a hammer blow; it was more or less hollow. A little gemmy treasure was disclosed within, one that had been hidden from every eye for much longer than human history altogether. Such earth stones—geodes, they are called—are often full of beautifully formed crystals. (Opening a geode is a thrilling little gamble; you can hardly know what you'll find until it is open. Some are disappointingly solid and dull.) Each crystal surface inside is a tiny natural facet, smooth, gleaming, angular, trim, but evidently in no way shaped or polished under the deft hand of any lapidary.

Some form resides naturally within mineral matter that wants to express itself in an elegant geometry. Ever since humans dwelt in caves,

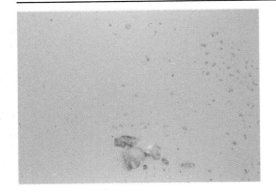

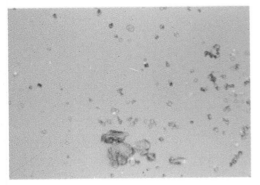

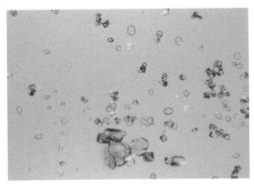

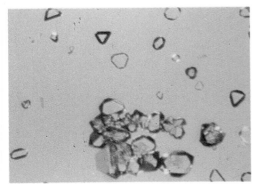

Mineralogist's time machine: specimens in seconds. A series of well-formed gems of ordinary alum assembled themselves out of a clear solution while we waited, to be seen only under magnification.

they have been caught by the unexpected order of these natural crystal forms. That order is another internal clue that we find almost wherever we look within the mineral kingdom—and do not forget the snowflake.

The growth of big mineral crystals in nature is slow and unseen. But if we can learn to enjoy small crystals, their assembly takes very little time. Can we give up size to gain time? The clever way is to watch crystals grow under the microscope; they will appear large enough to admire while they are still so small in reality that they would go unnoticed in the rocks. It takes some experience to prepare the right conditions.

Two different substances we tried grew beautifully as one watched. A hedgerow of pure silver metal emerged from a clear solution, its minute fronds faceted and mirrorlike. We do not usually regard metals as crystalline, though they almost always are. Their crystals are invisibly small and often distorted by working the metal.

From another water-clear solution there grew little transparent crystals of alum. Within half an hour the crystals had grown to the size of a grain of table salt. They would be museum specimens were we small enough to collect on their scale. The feeling of the inner necessity of

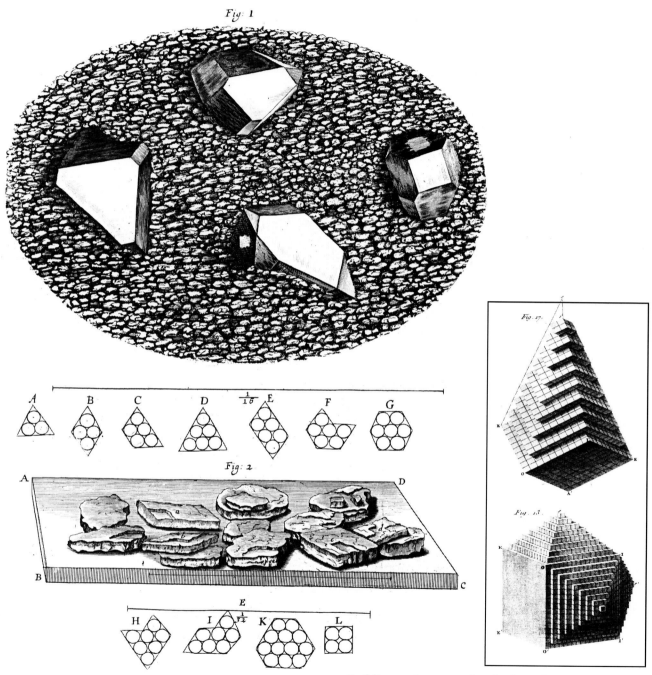

Crystals under Robert Hooke's microscope. Billiard balls rack up into regular figures with only a few kinds of corners. These flat arrays of "globules" were mused upon as crystal models in 1665 by the inventive Hooke, who did not go on to three dimensions.

Building with strange bricks: how the spatial forms of crystals could all be built out of little identical blocks was demonstrated about 1800 by a Paris mineralogist. His drawings include a stepped octahedron built naturally out of tiny cubes, a model for the pyrite crystals we had seen. (A cube of cubes is easy!)

A diamond in the rough is diamond-shaped; this gem grew neatly octahedral without the attention of any diamond-cutter.

Iron pyrite crystals take many forms. A cubical urge lies within them all; the cube appears at the left, and a careful look at the right-hand cluster finds many octahedral points.

Cubes from the saltshaker. Some table salt is neatly cubical under the lens.

form is inescapable as the geometrical facets form one after another out of clear fluid.

The commonplace mineral called iron pyrite often grows naturally into strikingly well-formed cubes. Often we find pyrite crystals that are not cubes, but eight-sided solids, octahedra; but in fact, their six corners point precisely at the faces of an imaginary cube that might be drawn to enclose them. They are part of one family, whether they have the one shape or the other. The cubical order is internal.

There is cubical order within humbler crystals yet; table salt can be beautifully cubical, grain after grain. Its cubes are too small to see without magnification.

Internal form, judged more by study of angles between crystal faces than by overall shapes, persists through all sizes and all colors to divide minerals most satisfactorily into families of form. By about 1800 the geometricians came to see how the special forms crystals took could all be built up by the superposition of tiny modular bricks, each shape of module determining a family of forms. Again this was not yet atoms, but it was coming close.

Among such common minerals it was hard to resist placing one rare one, a natural diamond. It does not have the scores of cut facets seen in the advertisements. That complex shape of the engagement stone has been carefully ground by the lapidary. The diamond we saw had grown naturally with the eight sides of

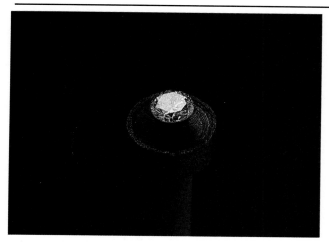

The cut gem diamond ready for the fiery furnace.

The diamond glows
hotter and hotter.

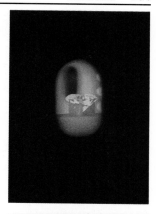

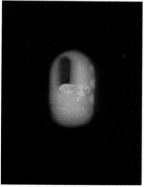

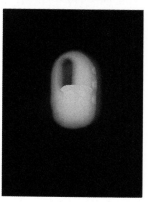

an octahedron. A diamond is diamond-shaped, as it should be.

ANOTHER MUCH STRANGER FORM OF COOKery can underline the internal geometry of matter. Once again we follow an old path. We secured a gem diamond. Then we found a helpful and expert MIT laboratory that had a vacuum furnace, one that can produce a high temperature within a sealed space from which all the air has been pumped out. After a number of trials with smaller and less valuable diamonds, we put our diamond in to cook, all by itself in the vacuum, white-hot above 3,000° Fahrenheit for about three hours. Once it cooled down, we opened the seals to remove what was left with a sense of loss.

All that is left of our gem diamond is here on a small plate. The black residue much resembles the charcoal we got from foodstuff. (When Julia Child first saw it, she said it looked "like a miniature burned soufflé, a very sad sight.") I tried a crude comparison: first, I drew a streak on paper with diamond charcoal; then another using instead a bit of the charred string bean. The streaks were much the same.

There was no air present while the diamond was hot. Had air been admitted, the white-hot diamond would not simply have charred black, but would have burned away into invisible gaseous carbon dioxide, just as the carbon in the

All that was left of a hard bright gem.

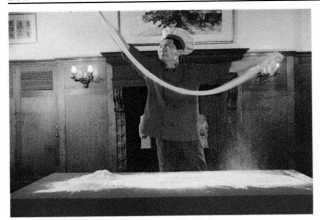

Making many noodles: fold and fold again.

food we eat is first oxidized internally, then exhaled as carbon dioxide in our breath. With care you can recover almost the full weight of the diamond as charcoal. (Gem diamonds, they told me, run better than 99 and 44/100 percent carbon!)

But there is a important difference between diamond and charcoal. They are not at all the same thing, as the most casual comparison shopping will convince you. What can that difference be? The atoms left behind are nearly all carbon atoms, and very little was lost as the diamond charred. The difference must have been in the geometrical arrangement of those carbon atoms in space. Diamond is one special pattern of carbon atoms; sufficiently heated, they regroup to form another pattern, the more familiar one of charcoal. That same stuff was left behind when the green beans—a material much more complex than diamond, containing many other sorts of atoms as well—were broken down by oven heat until their least volatile constituent, carbon, remained as charcoal.

Size

THE CLUE FROM CRYSTAL FORM IS WIDELY valid. Atoms in matter are able to array themselves in specific patterns in space. Surely we should try to find the geometrical volume one atom occupies, what we might regard as its size. That will be our next venture.

To find the atom we must be prepared to subdivide matter very much indeed, to cut and cut again until we reach the uncuttable. A most attractive way is to try that division according to a geometrical progression rather than by a simple arithmetical scheme. Instead of simply splitting off the smallest piece we can, can we try simply halving the whole sample and then halving and halving and halving it yet again, until we reach the small size and incredibly large population count of the atoms themselves?

A fascinating advance in that direction—to be sure, without atoms at all in mind—has long been practiced in the high art of the professional Chinese chef. We dropped in on such a virtuoso splitter of matter.

Chef Mark kneaded high-gluten white flour carefully along with the other ingredients of noodle dough in correct proportion: three cups of flour, half as much water, one-quarter teaspoon each of salt and baking soda. He vigorously swung and stretched the lump of dough out into a heavy single strand the length of his full two-arm span. Then he folded that long thick strand in half, and pulled the dough out again to its initial length, so that two thinner strands now passed from one hand to the other. Repeat, repeat, repeat…

CHEF MARK: Hello, everybody. I am the chef of the Dragon House in Wildwood,

New Jersey. Today I will make the kind of noodles called *so*. Make the dough strong and smooth, keep the dough smooth and strong, and you will have noodles on the table.

Fold one time: the dough becomes two noodles. Two times, it becomes four noodles. Three, four times…ten, eleven, now twelve doublings, or four thousand and ninety-six noodles.

Wow! Almost five feet long, they are called dragon's beard noodles—very fine, like a human hair.

In two minutes, Chef Mark had drawn out four miles of fine noodles. (They were really two or three times as thick as human hair.) Legendary chefs of the past have gone to thirteen doublings, while experienced home noodle makers can complete eight or ten. But consider that if Chef Mark had continued the doubling, it would take only thirty-five more steps of doubling, six more minutes' work, before he would have reached what we know as atomic size. Of course, the actual procedure would fail long before that idealized atomic limit is reached. (Dough is not at all a simple substance, and it is probably the evaporation of the water from the fine strands that sets the real physical limit somewhere close to a dozen doublings.)

The tantalizing nature of the doubling process is that the subdivision is so rapid. Some forty-six doublings would make noodles of true atomic fineness, in principle. But note that such an incredible feat would produce not a mere few miles of dragon's beard, but noodles long enough to stretch to Pluto and beyond!

Gold is beaten thin indeed; a gold leaf is two or three microinches thick, a thousand times thinner than a human hair. But no handwork can divide material down to the size of an atom, though by simple means that arise automatically in natural processes we can come remarkably close.

ON A SUMMER DAY OF VERY LIGHT BREEZE WE carefully placed a small spoonful of the best olive oil—biodegradable, and in fact delicious—onto the smooth surface of a pond in the park. From a bridge I watched the oil slick grow and spread out from where the oil entered until it had covered a large patch of pond. It can be seen well only because the slick is so nearly mirror-smooth that the reflections within it are clearly seen; elsewhere the water surface was rippled in the light breeze, and the surface reflections jumbled and confused. (The rainbow layer of oil one often sees on street puddles is in contrast quite thick, and was to be observed in our experiment only during the first few seconds of spreading.)

I could follow the surface of the slick as it spread from the spoonful of oil that we had put in. (Some ducks swam here and there shedding

The rainbow colors of oil on water arise when the layer is a tenth of the thickness of paper. It will spread much thinner.

Our oil slick spreads, made visible by the near-mirror smoothness of the water it covers. The ducks spread some oil, too.

oil, in competition with us.) The film spread at first fast and then very slowly, until it seemed to spread no more. I reckoned that the area it finally covered measured about ten yards by twenty yards, or two hundred square yards. Now, a square yard is three feet by three feet: call that ten square feet. (A physicist's arithmetic should not be more precise than his eyeball measurements.) We had made two thousand square feet of film.

From the measurements of the kitchen quarter-teaspoon we used we can figure that the oil increased in area by half a million times. The volume of oil stays fixed as the layer spreads. So the oil layer must have thinned in the same ratio that its area grew, down from the starting depth of about one-eighth of an inch.

The film we could see as the spreading stopped was on average ten or twenty times thinner than the gold-beaters make their airy leaf. But it is not yet at the size of one single atom. Oil is not made of a single sort of atom; it is a complicated compound. Its modules consist of some little cluster of atoms. That little cluster, whatever it may be, is not bigger than a five-millionth part of an inch, a fifth part of a microinch.

This is no new argument; it was put forward close to a hundred years ago from work indoors in the lab. But more than two hundred years

BROWNIAN MOTION

Common sense associates motion with animate life, in the absence of evident external influences like wind or wave or the pull of gravity. Under the microscope tiny organisms clearly move on their own within droplets of water. Robert Brown was a London botanist of wide interests and distinction whose work at the microscope had made clear the existence of a special body within living cells to which he first gave the name "nucleus." In 1827 Brown noticed that within the hollow grains of a certain microscopic pollen very tiny particles indeed were in steady haphazard motion.

First, Brown made sure their motion was not due to currents or other disturbances. Could those particles be alive? It seemed that they were not; boiling made no difference. Anything he could make into a powder fine enough to stay suspended in water moved in the same strange way under his microscope, even ground-up bits of the Great Pyramid!

The discovery of this enigmatic motion was a landmark; it has been called Brownian motion ever since. No one understood its nature and origin for years, until the early part of this century, when the new successes of the atomic theory led both Albert Einstein and a young Polish theorist, Marian Smoluchowski, to offer independently a full explanation.

Water, like every liquid and gas, is made of atomic clusters in constant random motion. They are far too small to see. But if a fine particle just large enough to be seen with the microscope is suspended in any fluid, the random jostling that the extremely light particle receives on all sides from the atomic motion will move it randomly in turn. Larger particles are too massive to respond noticeably to the molecular trembling. All the details of the motion, a truly random walk, can be predicted statistically with success.

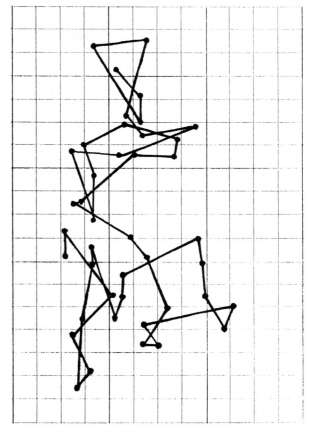

The random walk of Brownian motion. This aimless wandering path records the actual positions of a tiny suspended particle seen in the microscope, steadily batted around by the invisible molecules of water. Its location was noted every thirty seconds. The straight segments that join the observed positions are physically meaningless; the path the particle took betweentimes was just as jagged as the path drawn, random still but on a smaller scale.

In our electronic age the electrical equivalent of Brownian motion is even more important. The current carriers in every conductor feel the random atomic motion. That imposes a tiny random signal upon every current; when we amplify greatly, as we so often do, the motion is a source of random electrical noise for which there is no relief save to cool the input structures. Atomic motion is the slower the lower the temperature.

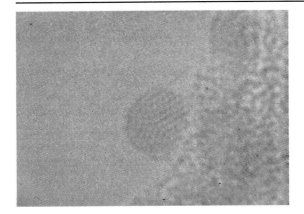

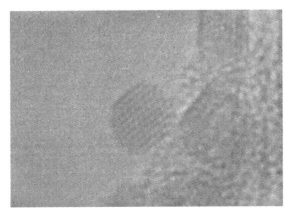

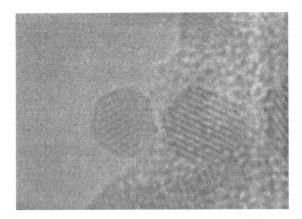

Atoms on parade. Video images of tiny gold crystals in motion, some with as few as a thousand or two gold atoms, their ordered array clear though shifting. The magnification overall is about ten million times.

ago Benjamin Franklin became interested, not in atoms, but in the more practical matter of calming troubled waters with oil. He found then that it took about two teaspoons of oil to calm one acre of water surface, pretty close to the value we ourselves found at the pond. The ability to measure some limit on the size of an atom by this understandable technique is a strong argument for the existence of a size scale for the atom.

Nor could these atoms be very much smaller than a hundredth part of a microinch or so. If they were very much smaller, if the oil film could get a great deal thinner, it would have spread almost indefinitely. Yet it did not do that, either for us or for Ben Franklin two hundred years ago.

AT ARIZONA STATE UNIVERSITY IN TEMPE there is a transmission electron microscope with a beam strong enough to allow repeating pictures at video speed. It is being used to make atomic video movies at enormous magnifications up to tens of millions. Still photographs of arrays of heavy atoms like gold are not really new, but real-time movies like the ones from Tempe are unique. We can compare these modern atomic images with what we have already found about the size of atoms.

The forms we could examine were tiny crystals of gold, whose heavy atoms provide strong contrast in this kind of microscope. They are

 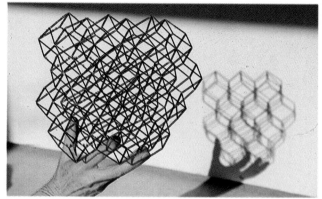

Simple shadows of an intricate array. Some help may be found in a shadow game, meant not as a precise model but as an analogy. What we produced was a crisp shadow of a wire latticework. As the wire frame turned and twisted in the beam of light, the shadows caught something of what we had seen in the video microscope among the moving atomic arrays of gold crystals. There is a strong sense of order almost at hand. Every once in a while, as if by chance, an unexpectedly simple view appeared as a whole column of shapes aligned so that they were all shadowed as one.

spread out on a very thin film of carbon, whose individual atoms are not very different in size from the gold atoms, but much lighter. Lighter atoms do not produce visible shadows because they are more or less transparent to the electron beam.

The whole screen images a tiny area of the sample. To use the controls of this device to scan steadily for an hour in a straight line would carry the view only across the width of a pinhead. The particles of gold that are seen here are crystals so small that the entire crystal would remain quite unseen in any ordinary light microscope.

But they are visible to this instrument in atomic detail, displaying the kind of ordered geometry of crystalline atoms that we expect. It is fairly easy to measure the size of the gold atoms in their ranks. As one watched the video images, the small crystals shifted and rotated; now and then some portions would even jump a little, or seem even to fly away. Sometimes we would look into a complex a hundred atoms or so thick, occasionally only at a thin edge a few atoms deep. All that swift mobility is not entirely understood; the motions are being sampled at normal TV speed, very ill-suited to atomic motion. These atoms are moving rapidly in part under disturbance by the fierce radiation of the probing electron beam.

David Smith's wonderful electron microscope can image so tiny a speck of gold that we see its atoms, not one by one, but arrayed in regular order, in a paradelike motion that recalls some musical films of the 1930s. As the gold crystals move it is not always easy to interpret the complicated shifting forms that are the atomic shadows.

Once we had caught an endwise view down a row of atomic shadows—the dark holes in the video images—it seemed easy to judge the spatial separation of the atomic layers aright. That turned out to be about a hundredth part of a microinch, a hundred-millionth of an inch, two dozen times less than the minimum thickness we found so easily for an oil film upon water. We knew that the oil layer must have been larger than a single atom, because the smallest unit of oil is a cluster of many atoms. (The chemists have long used the reactions of such an oil to characterize its building block, its module of structure, as chiefly a long string of a dozen or two linked carbon atoms.) But we expected also that we had not overestimated the atomic size by a huge number. The transmission electron microscope concurs.

The size of atoms is confirmed daily by many modern instruments, though by few as directly as the video electron microscope that Dr. Smith has put to such vivid use. The atoms of ordinary matter turn out to differ relatively little in size from one element to another. It is a little surprising that the atoms of gold, nearly twenty times as heavy as the atoms of carbon,

GOLD

THE GOLDSMITH'S ART IS ANCIENT and cosmopolitan, subtle and varied. Yet it has a unity of fulfillment that rests both on the universality of human genius and spirit and on the specific qualities of that atomic lattice. We exhibit half a dozen superb examples.

Five hundred years before Abraham left Ur of the Chaldees, the fluted cup had been buried there to provide royal drink in the afterworld. Its flowing form in hammered sheet gold speaks of easy mastery of the most workable of metals, achieved more than 4500 years ago.

The art of granulation had been lost for a long time, but this half-inch gold bead with about 2,000 granules on its surface is contemporary work: the metal has its adepts still.

The animal designs on this pin are wrought in tiny sparkling gold beads, called granulation, all set in place by a steady hand, but made and fastened elegantly and permanently through the mastery of gold in fire. Control was needed both for rounding the spheres and for the ingenious procedure that welds minute bead to surface. The piece is Etruscan, where the technique grew twenty-five centuries back.

Goldwork in the New World owed nothing to the Old. The hollow jaguar from ancient Peru, mediator between humans and the cayman god of the sky, is composed of a dozen shaped pieces, each of uniform hand-hammered sheet. The joining employs a variety of fusion and soldering techniques all under knowing control. The piece is around 2500 years old, one among seven like it that once made a single sacred ornament, and now are held and admired over three continents.

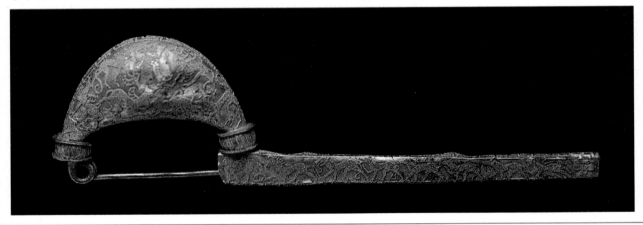

The small costumed personage from Colombia is certainly more than a thousand years old. It is cast from gold, and the subtle colors on its varied surface were laid down by treatments that enriched the outer layers in gold content. The jaguar, too, had its surface enriched toward purer gold.

As bright as the day they were made some five centuries ago, the sun images foretell and describe partial eclipses. The half-hidden solar disks gleam in beaten gold leaf on the vellum pages of a manuscript almanac made for personal use.

are still about the same diameter. Each experiment that has pointed to the atom has pointed to the same sort of modular atom, all more or less the same in size, marked by an implied geometrical simplicity of form and pattern at the heart of matter, and present everywhere in a teeming multitude of copies in incessant motion.

The oldest experience we could follow toward the atom was that of the artisans. Workers in gold found through the long practice of their trade that the properties they desired in gold—its glitter, its color, its malleability, its permanence—grew more and more sharply defined as they learned better how to purify the metal. In the end many craftsmen became convinced from experience that pure gold is one single stuff that always behaves the same way, whatever the mine that offered the metal, whatever century had mined it.

Most of the early philosophers, like the celebrated alchemists who followed them, had taken a different track. They thought that a searching analysis of matter would isolate pure properties rather than pure materials. The glitter itself was the end to seek, not the most glittering among metals. It was less important to win specific substances. Some said that it is air, water, and earth that make all matter, along with fire. Those ancient categories, if not taken too literally, describe matter less well than they nicely fit the properties held by matter, for they

Flame colors: a hot, faintly blue gas flame is brightened by the crimson of strontium, the green of barium, the blue of copper, the lavender of rubidium, the yellow of sodium. (The eye, the emulsion, and the printer's inks do not always concur on the subtleties of color. These are at best approximations; each instrument has its own response.)

The colored flame rises above the hot wire and the pale blue glow of the burning gas. That wire had been dipped into the test substance; a few grains cause the color.

Color is a clue; the flame and the electrical discharge match in color. The element rubidium is present in both glowing gases.

recall gas, liquid, and solid states more than they suggest concrete examples like gold, mercury, or oxygen. (Fire surely was taken to represent the genuine transforming role of heat.)

I believe the old philosophers guessed wrong. The alchemists never did find their way to the properties of gold starting from baser metals. Rather it was the master craftsmen, experienced and intuitive, who were on the right track. Properties best follow particular substances, and the bundle of properties you really want could come only with a well-defined single substance.

There were, of course, some philosophers of atomism, too, early heralds of the idea in Greece and in Rome. But those who first came up with strong evidence for the atom, even though they were never to see atoms at all and could neither count nor measure them, were the brilliant pioneer chemists of the nineteenth century.

They studied the whole web of chemical reactions, a tangle of relationships far more elaborate than the few we have touched upon. If matter is a set of a few kinds of atoms that link and unlink in myriads of tiny clusters arranged in space, that model would explain all the reactions they found. They became convinced that their everyday work was done with atoms in bulk, without ever seeing any image of an atom, or ever knowing for sure its size. The logical structure of what they had found, largely a geometrical view, made sense of so

many chemical processes observed only at human scale that they were led to embrace a model. They conceived of an unseen but universal grainy substructure for matter at extreme small scale, an architecture they could use predictively with unfailing success, although they could not demonstrate it directly.

It was to our twentieth century that the atom itself was first disclosed. Perhaps the most direct case was made by the spinthariscope, our spark viewer. One spark meant one fast atom stopped in its swift flight. I find myself still dazzled by this faintest of lights, a light that emanates from one of the rarest but most powerful of atomic processes, radioactivity.

Taken together, all those experiences are compelling. By now we may accept beyond any reasonable doubt that a modest menagerie of atomic species arranged in an outrageous variety of geometrical patterns at many scales do make up essentially all the matter we know. Every substance is but some populous arrangement of atoms continually moving in space.

But for the inward behavior of the atoms themselves, we must pursue thought and experience beyond ordinary substance into the shining realm of fire and light.

The Quantum Atom

FIREWORKS ARE AN OLD CRAFT, TRANS-formed by the growth of our understanding of substance. The exciting flight of the skyrocket,

its train of yellow sparks, and the sharp report of its burst are as old as gunpowder, but the rocket's green or crimson glare—that is a gift to the public and the craftsmen from the chemists of the eighteenth century. Fireworks are only the showiest realization of such unexpected light, the product of specific substances incorporated in the glowing materials. Colors well beyond the reds, oranges, and yellows of any fireplace can be added to an ordinary gas flame to transcend its faint blue envelope and yellow core: lavender, green, bright yellow…All you need do is supply the hot flame with a tiny pinch of a particular substance, a little of one or another element, sometimes rare ones, sometimes common. Strontium compounds in general give crimson flame, rubidium salts lavender, and any small amount of sodium salts produces a strong yellow. It is not hard to build up a lore of flame colors closely akin to the more elaborate recipes of the fireworks maker.

Have you ever ignited an ordinary highway flare? It is a somewhat exciting if long-burning kind of firework, whose characteristic red flame signals danger on the highway. The eye can pick up that bright color from quite a distance. Yet the crimson hue is the sign of the deliberate choice of composition of the burning substance: a compound of the element strontium is present. So the eye can judge some materials even from a great distance. That is remarkable enough.

But we can do better than that. If we place

Highway flares seen with an optical grating over the lens. The grating spreads to the right of each flare a set of images of that flare, lined up color by single color. Each set is a spectrum of its flare.

before the single eye of the camera a device called an optical grating, a piece of plastic crossed by an invisibly fine ruling of parallel grooves, we can see something quite remarkable. The grating can be slipped directly in front of a photographic lens (or held before the eye).

Then the single flame shows up with many visible partners beside it in the air, each with the same form as the one true flame of which they are obedient images, moving with the flame as it moves. They form a series of ordered images in different colors, all clearly coming from the flare itself.

The grating has separated light on its way into eye or camera according to some intrinsic quality of the light, the distinct images spread out in a row. In a direct view all the eye or the camera sees is the constituent colors of the flare one on top of another as a single image, an image in light whose overall color the eye can judge only as a kind of resultant sum. Here the flare color has been taken apart, as water can be taken apart chemically into its two elements. The analyzed colors now appear distinct, perhaps plainer to see.

Raindrops have performed a similar analysis for a long time, breaking the white appearance of sunlight into the hues of the rainbow. A glass prism does the same task. The spread of analyzed color seen from any source of light is called the spectrum of that source. These devices that can form spectral spreads out of

any light source are evidently extensions of the color sense of the normal human eye.

We could have chosen to look at a green highway flare. It has a different chemical recipe: the fireworks maker will tell you that in the green flare a salt of the element barium has replaced the strontium. Now several green images are found along the spectrum of the green flare that are not present for the red flare. The spectrum makes possible an objective judgment of those differences in color, merely by noting the positions of the images along the spectrum (they would be recognizable by position even to the color-blind or in black-and-white photography).

The eye sees a bright yellow flame whenever any compound of sodium is present. But in a carefully spread out spectrum it can be seen that all the yellow color is conferred by one single image of a special yellow, at one specific location. The overall color of a flame to the eye is a crude assessment of what light was present; you could find the sodium yellow in light that offered the eye no hint of yellow, because other stronger colors masked the yellow. The spectrum of a gas flame colored by added sodium shows almost nothing except for one image in the yellow.

In order to see more detail within the arrangement of colored images, the spectral device can be fed its light, not as an image of the whole flame, but by sending the light in through a narrow slit. The image of that slit will

Lightning records its spectrum. The image of any source in each of its spectral colors has the same form; a zigzag source has zigzag images.

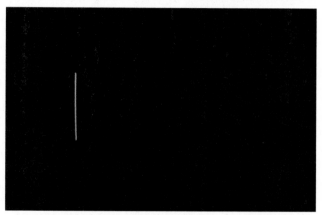

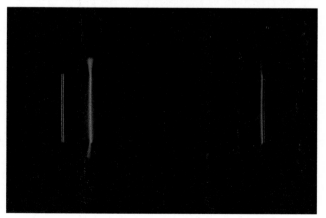

appear in all the colors present in the light as did the flame itself. But now overlap of the images from neighboring colors will be minimized. A lightning bolt is itself so narrow a source that it needs no slit to show a spectrum of distinct close-lying colored images. The spectrum of a sodium flame seen through a slit is reduced to one or maybe two very close yellow lines, an image of that slit in one color position. It is the bony skeleton of a rainbow.

By the mid-nineteenth century such diagnostic spectra had become an analytical tool in the chemical laboratories; the spectroscope could identify the lines of colors associated with very small amounts of one particular element in the flame. A dozen new elements were first found as enigmatic spectral patterns, and only then chemically isolated as visible samples.

Not all elements could be found that way, nor did all light sources show interesting features. It turns out that light from glowing gases, as in a flame or a neon sign, has distinct line patterns; when instead it is a solid that glows, like the filament within an incandescent bulb, or the white-coated tube of a fluorescent lamp, a bland rainbow appears. But gaseous flames (along with sparks and other bright electrical discharges in gases) typically were marked by patterns of lines that could be identified by

The bones of a rainbow. A laboratory spectrum is a set of colored lines, images of a narrow straight line source, chosen only because that form reduces overlap. Here is the rainbow band from white light, featureless but colorful, and below it a narrow-slit spectrum of glowing sodium vapor, all the light appearing in one yellow line. The bottom spectrum is one of mercury, its lines similar to those of blue-green mercury vapor streetlights.

Lights in the wide world show spectra as well as flares do. The camera and its grating were taken to a brightly lit lot: a jumble of rainbow bands. We will see such a pattern again.

Invisible lines of mercury vapor. In the upper spectrum the mercury lines were projected on a white wall. In the lower one you can see a framed receiving screen in place as well. That screen is ultraviolet sensitive. On it the two violet lines above are supplemented by a third line, one that looks yellowish though it is far to the right, beyond the violet positions. The eye and the film cannot see that line directly, but it is transformed by the screen pigment into a yellow visible to both. The position is what counts!

direct comparison with light from sources prepared element by element in the lab.

Such a known pattern of lines amounts to a fingerprint of an element. The set of colors of the lines was usually noted down by some impersonal measurement of position along the spectrum. They defined the pattern. Of course, each spectroscope has its own spread of color; comparison and calibration are not difficult. The spectrum seemed as much a property of the atoms of the unusual metal mercury, for instance, as its silvery fluid form when it is pure. But the fingerprint was usually present even when mercury was added to a flame only as a compound, and never present when there were no atoms of mercury. These days the spectrum of mercury vapor is not hard to find; the light from mercury vapor alone is an overall bluish color, both queer and familiar. Many bright street lamps emit such light, the strong glow of electrically excited mercury vapor within the lamp. There are half a dozen lines to be seen in the visible spectrum from mercury vapor.

Soon enough there were lists of measured

A bee's-eye view. The blue image shows a blossom of evening primrose photographed in ultraviolet light. A dark spot surrounds the center where nectar is to be had. In the photo taken in human colors, no such marking is seen. Surely the bees can see in the ultraviolet to use that special signal. They have optical instrumentation unlike our own.

positions of thousands of lines catalogued by the elements that gave rise to each line; even superimposed "fingerprints" could then be disentangled to identify many of the elements within some glowing source. Evidently it is the atoms of the element that emit the light. The specific patterns argue that the atoms of a given element are capable of emitting only those observed lines.

The analysis of light color proved a working substitute for the still-impossible dream of voyaging to a distant star to collect a sample. We were now in possession of a way to gather the recipe of the universe, a heady idea. Spectroscopes could analyze the distant glowing gas of the sun and the stars; the elements found there were the familiar elements of earth. The new element helium (it is named for the sun) was first known only as the origin of certain yellow lines in the spectrum of the edge of the sun, lines unknown on earth. They were eventually shown to arise from a new gaseous element that turned up in certain rare mineral samples. Helium was far too costly for toy balloons until it was later recognized in large amounts within the natural gas from a few fields in the United States and Canada. (One discovery was made when the gas first intended to light the municipal celebration of a new Kansas find simply refused to burn: too much helium.)

All the spectral features lie somewhere in the color band of sunshine, among the colors of the rainbow. The eye reports a range of colors from red to violet. But when sunlight is spread out by prism or grating, it is clear that beyond the end of visible light in the violet there is invisible light, physically continuous with the rest of the spectrum, unseen only because it is of a color not normally perceived by the human eye. We call it ultraviolet.

The spectrum of mercury in fact has its strongest lines there in the ultraviolet, well past the last violet line that is normally visible in the spectrum of a mercury vapor lamp. Any of a large number of chemical substances can convert some of the ultraviolet light to a visible color; a card coated with such a fluorescent paint placed in the mercury spectrum beyond the violet limit makes visible—in another color—the mercury lines in the ultraviolet. The farther ultraviolet will not pass through ordinary glass, and extreme ultraviolet light does not even penetrate the air. (The usual fluorescent tube is a mercury vapor lamp coated with a fluorescent screen to capture and render visible the flood of ultraviolet light from the glowing bluish vapor within.)

The atoms that had strong lines within the visible range of color were detectable from the first. But lines outside the visible color range could be found only by some special means of detection. Photographic film is quite sensitive to the ultraviolet, for example; color film usually reports ultraviolet as white. Here is one reason for the puzzle that in early work many elements did not appear to yield any lines at all; their lines were present all right, but the investigators missed them because they lay outside the range of human color vision.

IT IS TIME TO RECALL WHAT WE HAVE learned about atoms along the varied paths of experience from gold-beating to spectra. Atoms seem real; they come in huge but finite crowds of little distinct entities, sometimes arrayed with their fellows, sometimes clustered in groups, always in motion. They can be sorted out into a modest number of classes, called the elements. We have counted atoms, measured their size, even watched their fast-moving video shadows enormously magnified.

One central question goes back to the days of the philosophers who speculated on the nature of substance and its changes. How is it that atoms, say of gold, can remain so uniform, so much the same, even though they all have different histories of incessant rearrangements and collisions? The old atomists felt that the atoms must be incredibly hard, unchanging bodies, so that they could endure like the noble metals themselves. They were incuttable, as the word *atom* says in Greek.

We know that is wrong. Atoms have a rich and complex structure, not at all hard to disassemble in part. It is enough to dissolve some elements in water or to vaporize them in a flame to begin to disturb the atomic architecture. The emission of characteristic light from atoms proclaims that some kind of internal structure is open to change.

It was the spectrum of light from the atom that led to the first understanding of its inner behavior. The course we will follow begins just before World War I; it was the recognition of the discontinuous behavior of the atom, the quantum theory.

By that time it was plain that every atom that emitted light had to gain energy somehow beforehand—usually from a collision of some sort—to lose it again in the energy of light. You had to put energy in to gain light: fuel, electrical power, chemical reagents. The energy of light was not stored in the atom. The colors of light that a given sort of atom emitted did not change; sodium yellow was always the same. The atoms did not wear out: they could gain and deliver energy over and over again.

A bright source of light gives more light energy than a faint one in a given time, certainly; that was nothing new. Consider for simplicity a source of one single color, say sodium yellow. A brighter yellow light is one in which more sodium atoms are acting to emit light. Yet each of the identical sodium atoms made its own fixed contribution to that total, as it was excited and then emitted its energy, usually over and over again. (An atom that has gained energy is said to have been excited.)

The collisions energized the atom; it gave out light of one color, but the energy in that tiny parcel of emitted light had to match the energy the atom had picked up. The light from one atom took away energy according to its color. That is checked by noting that light of one color always gives a fixed amount of energy to any atom that has absorbed it.

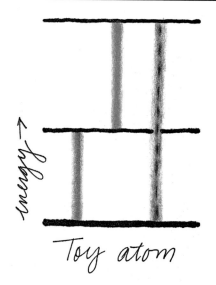

Toy atom

Energy states of a toy atom. No real atom is as simple as this toy atom, with only three levels. Their differences correspond to the line colors shown. Drawn to the right is its three-line spectrum, with one line in the ultraviolet. The sum of the measured red and green line energies would match the energy of the ultraviolet line; finding that agreement would strongly support the theory.

The sharp bony spectrum tantalized the physicists. The sodium atom sang a one-note tune, to use a musical comparison. Maybe there was a single structure within it that picked up a fixed energy and then gave it out, always matched to what came in, as energy changes must always balance. Then more complex spectra would mean many such states within an atom, reminding one of the strings of a piano.

It was plausible, but it never worked. No such scheme agreed with any atomic spectrum. The key was finally turned by Niels Bohr around 1913. Each atom indeed has a series of sharp energy levels or steps or rungs. They are the only energy values that it can hold. Whenever an atom radiates energy as light, the light energy that goes off is of course lost to that atom. The atom must then take on another state of lower energy, which must be one of its characteristic levels.

Bohr found that every spectral line has the color it has—or the corresponding energy—because the energy the light takes off from each atom matches the energy *difference* between *two* levels of that atom.

Bohr's scheme, demanding initial and final levels for every line, worked superbly. He began with a theory of the spectrum of hydrogen, the simplest of all atoms. Its success led to the general result.

LET US DISCUSS NOT A REAL ATOM BUT A SIMplified model for study. Suppose our atom has only three characteristic energy levels. The diagram shows those three levels entered on a vertical scale of increasing energy.

What spectrum is emitted by our toy atom? Imagine a sample of the toy element; call it by the symbol T. Plenty of atoms are present, huge numbers within almost any sample of T. Of course, in a bright source of light we see emissions from many atoms in a given time; in a weak source, far fewer atoms manage to emit their light.

Every atom of T has the same set of three energy states for its internal energy. The light from the flame can show only three lines in the T spectrum. One line comes from any atom that jumps from the top state to the bottom, another from a jump down from the top to the middle level, a third from a jump out of the middle to the lowest. The energy *differences* of the levels give each line energy, and hence the position along the spectrum for each line. The atomic fingerprint of T is plain. Now the energies of the three lines, which can be calculated from the colors of the lines emitted, are related. The energy of the most energetic line must equal the sum of the energies of the other two! That is the strongest of surprises; it is very different from lines that could be assigned any value a state might have.

The surgical apparatus. The lamp is housed in a device that can control electrically the energy available to any of the atoms in the glowing gas. The light from the lamp passes through the small box with a dial which is able to pass light on color by color as the dial is turned to form the spectrum. The light is received not by eye or camera but by a photo-electric detector that sees all these colors and records what it sees.

We can devise a different test. Suppose we treat our sample of the element carefully, not exciting the atoms randomly and vigorously, but adding energy under tight control. If we add enough total energy so that many atoms are excited, we will get enough light to measure. But if we control the maximum collision energy for any one atom, we can arrange that among many moderate collisions none will bring in energy enough to send an atom all the way from the lowest state to the top state. Then atoms might be made to reach only the middle level. Even if some atoms were for a while in the middle level, the moderate collisions could not send them to the top. Then the sample can yield only one line of its three-line spectrum. That three-line fingerprint is the result of a physical process, not simply a mysterious imprint.

The characteristic energy states for any atom plotted against their energy form a diagram that we may call the quantum ladder of the atom, a more graphic term than the formal title: energy-level diagram. The quantum ladder for any atom type is a kind of map of spectral behavior, and indeed of much more.

We have come to a physical theory of atomic behavior, called the quantum theory. That behavior is strange; quantum behavior is unlike much that we expect from everyday experience.

Consider further what the energy step means. Politely offer an atom some energy, yet not enough to reach its first step up. It will reject the offer, to recoil unchanged, as though it *were* a hard unyielding ball. Only once it is offered sufficient energy to jump up one step or more can internal change occur. In between the internal energy levels, within those gaps of energy, atoms do act as though they were unchangeable.

The quantum ladder leads to a world of specific and lasting atomic properties, which are nevertheless not eternal. That is our world, the real world, where all gold glitters, yet it can be melted, and all diamonds are hard, and yet can be burned. It is a world of being and of becoming, a theory of matter rich enough to accommodate everything we have found so far.

THE EXPERIMENTS THAT FIRST EMPHASIZED this strange behavior to the physicists are not new. They go back to the years of the First World War and afterwards. But they are fundamental, and they fascinate, because with simple equipment right on the lab bench, using the experience of line spectra, we can observe quantum behavior within a sample of real atoms.

We tried a visible version of one of the most important experiments on the quantum nature of the atom, using the line spectrum of an atom. We could not procure any sample of the simple three-state element T; so we used mer-

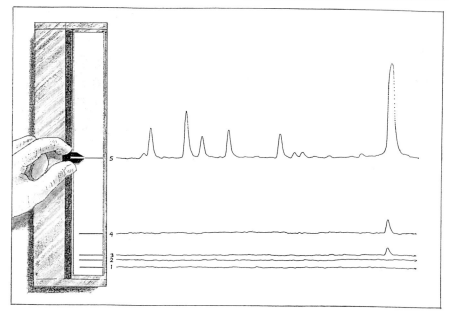

The spectrum dissected. This is a composite record of the five runs we made, each run showing a baseline drawn higher than the one before, to show how the exciting energy increased, run by run. The strong peak at the far right is the main ultraviolet line; the leftmost peak is the yellow line of the mercury atoms.

cury, for a variety of practical reasons. (The light source was a small bulb made of special ultraviolet transmitting glass. It had been manufactured for germicidal use inside a home refrigerator, now out of style.) The bulb gleamed an eerie mercury-vapor blue when it was on. It was housed in such a way that no stray light could enter the optical system, only light direct from the bulb.

The light entered a neat little optical device that filtered the overall light rather than spreading it as the grating had done. It amounts to the same effect; the filter could be adjusted to pass along a narrow band of color around any color at which it was set. An electronic detector was surrogate for my viewing eye; it was sensitive to visible and ultraviolet light alike, another example of how much a deepened experience owes to instrumental extensions of the human senses. Finally, a chart recorder drew a map of the spectrum seen by the detector. The spectral lines appeared as peaks on a graph as the pen moved horizontally across the chart, plotting at every color position as it was set by the filter. The height of the peak that the pen drew at each position marked the strength of the light signal there.

To produce light, we normally crowd in a lot of energy, letting the atoms collide as they will; but here we have another objective. We will exercise the tightest control we can on the energy of collision, on the vigor of the collisions among the stirred-up particles within the bulb. It is those collisions that transfer energy to the atoms that they may then emit as light. We control the collisional energy electrically (in fact, by radio power sent into the vapor within the bulb). There is an adjustable limit on how much energy is given to any colliding particle. The higher we set the knob, the higher the maximum energy available to any atom in a collision. The input energy scale is marked vertically, just as it is in our energy diagrams. The random motion from the temperature of the vapor in the bulb is small compared to the electrical excitation we have provided.

With this tabletop apparatus it took only a few minutes to approach the quantum nature of the atom. We made five energy settings, steadily increasing the maximum collisional energy available to the atoms of mercury in the bulb. At each setting of the excitation energy, the device plotted out the spectrum it saw.

First run: I set zero electrical excitation. There is always some small thermal excitation, unimportant here. The plotter moved across the calibrated chart, past the visible colors (the plot starts redward of the yellow line of mercury vapor) well into the ultraviolet. Nothing:

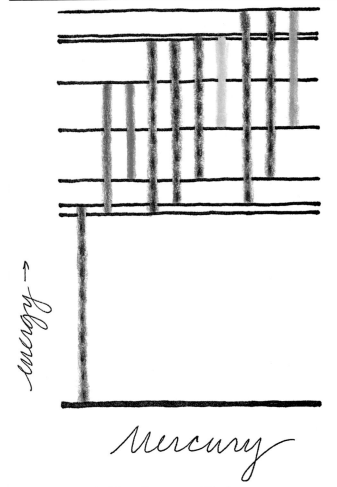

energy →

Mercury

The quantum ladder of mercury. The lower energy levels of the mercury atom are here drawn. It is this ladder whose rungs we crudely climbed in our experiment, sometimes a few at a time. The lines of the spectrum we found are drawn in their natural color. Ultraviolet lines, usually invisible to the human eye, are represented in a black-dashed purple.

good. We certainly expected nothing. The detector was quite sensitive, and a faint waviness in the baseline was the sign of random noise, always present. None of that faint wavy structure repeated from run to run; it is random and without meaning for us.

Second run: I put in a modest excitation energy by moving the input knob upward a bit. Again the device plotted what it saw. Nothing. The atoms were being offered some energy, but not enough of them accept it to excite any

detectable light at all. The atoms are waiting. There is an energy step we have not yet crossed.

Third run: More energy. The plotter drew again, but not a single line appears in the visible or beyond—until at last, toward the ultraviolet limit of our spectral range, a credible peak appears at a position familiar in the usual spectrum of mercury.

The scalpel of energy has dissected an atomic fingerprint; we have so prepared the atoms of mercury that only one line of its usual dozen-line fingerprint line pattern is to be seen.

Fourth run: More energy still, but no more lines; only one stronger peak at the same ultraviolet color as before. We had not crossed another step. The excitation energy is not under perfect control; the maximum offered has gone up, but also more atoms are receiving the smaller amount they need. The energy maximum is a little blunt, not a sharp edge at one precise value.

Fifth run: I added a lot more excitation energy. Now the plotter leapt up at once; all the visible lines are there, yellow, green, blue, violet, and several more in the ultraviolet; that first strong ultraviolet line remained. The excitation of the spectrum was about as complete as we have ever seen it. This is the familiar fingerprint of mercury.

Notice we got no light until we had crossed some sort of initial energy step. Then only one line was there. Then after a large increase, we found all the lines we expected in the spectrum. Perhaps if we had cut more finely in between, we might have found several layers of lines that would appear singly or in little groups. That takes time and a sharper energy probe.

We had used atomic spectra as though they

were something like fingerprints—either the full pattern was there, or nothing. The fingerprint of mercury is its spectrum. So people had thought very reasonably for a hundred years. But it is not so. The spectrum has a physical structure: we can dissect it in energy, to produce a line or a few lines at a time by controlling the energy given to the atoms.

That was the outcome of this experiment first done a lifetime ago. The widely spaced lines of every atomic spectrum seem to fit perfectly with the idea of a steplike response to offered energy. Even our experiment done with a coarse energy probe shows that much, I think. The best results go beautifully beyond what we have done.

There is not only the succession of lines we saw, but for each line there is a close fit between the measured energy gaps—the distance between the succeeding steps of the energy ladder—and the energy of the light given out in any jump from one atomic level down to another. The spectrum of light can be predicted without using any light at all, by electrical measurements of the input energy that excites the atoms. Once such a ladder is known, the behavior of the atom under all sorts of excitations can be at least partly foreseen.

The quantum ladder of the real atom, mercury, is not simple; our blunt probe could not tease out all the fine detail; it merged a few levels into a single response. More precise work using a sharp energy scalpel locates all the rungs of the ladder that determine the full, many-lined spectrum we can observe.

That deep discontinuity of the spaced levels, the gaps between the rungs neatly matched to every energy transfer, is the real meaning of the quantum jump. That jump is no proverb or metaphor or slogan, but in fact expresses what we see of the behavior of the discrete atoms of the elements that make up all the matter in the world.

A bouncing ball can take in or give out any available amount of internal energy within its limits, as a trombone can play any note. But

A CHEMIST'S LETTER

HERE IS A SOUVENIR FROM THE DAYS just before World War I when the quantum atom was new. It is a letter written by a blossoming young chemist, George de Hevesy, who worked then with Rutherford in Manchester, the same laboratory from which Niels Bohr had just returned to Copenhagen after a year's stay. Bohr had lately published his first quantum theory of the simplest atom, hydrogen, and its spectrum.

At a meeting in Vienna in 1913 the twenty-eight-year old chemist delightedly told Albert Einstein the latest news from Copenhagen. In those days Einstein was not yet an elderly symbol; he was a much admired physicist at thirty-four. Einstein himself was a major creator of the quantum theory. The younger man described to Einstein a recent confirmation of Bohr's wonderful scheme of energy levels, with its fundamental assumption that all spectral lines come not from one level but from energy differences between two levels. Bohr's theory of atomic structure included much more detail, and its predictions fitted closely the helium spectrum newly observed by a London experimenter, Alfred Fowler.

At once Hevesy wrote to Bohr. "When he [Einstein] heard this, he was extremely astonished and told me: 'Than the frequency of the light does not depend at all on the frequency of the electron....And this is an *enormous achievement*. The theory of Bohr must be then wright.'"

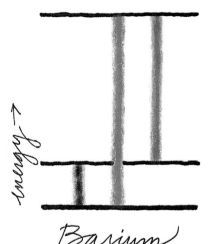

Quantum circus ladder. These three energy levels are the relevant part of the quantum ladder for the barium atom on center stage. The laser beams in the lab match the colors of two jumps shown; the other jump is unseen, in the infrared. The atom is put through its paces up and down these energy rungs as the experimenters direct.

consider: if a marimba had a million bars, do you think its tunes would sound any more limited to your ears than the trombone we regard as continuous?

The world is all quantum, I believe, but the rungs of every quantum ladder that describes the inner energy of a large-scale structure are so numerous and thus so close-set that they really cannot be told apart. Only at the tiny scale of molecular and atomic structure and below are the energy steps spaced widely enough to make a strong discontinuous mark on events.

AT THE UNIVERSITY OF WASHINGTON IN Seattle, Professor Hans Dehmelt and his colleagues have become talented ringmasters of a well-trained atomic circus. Their most spectacular act is persuading a single atom to stand still all by itself so that they might take a clear picture.

The most conspicuous features of this lab are the two laser beams, one red, one blue-green, that stab the darkness. Those lasers are carefully tuned almost to match the energy gaps between relevant states of the type of atom they will use. The diagram reproduces the quantum ladder in question, the jumps marked by color. That diagram is a key guide to the behavior of their trained atoms.

Let the ringmaster introduce his celebrated act, candidate for the greatest atomic show on earth.

DEHMELT: An ion is an atom in a particular kind of state: it has lost at least one of its electrons and so it has an electric charge. An ion in a trap is similar to a marble in a bowl. If you give the marble in the bowl a push, it will roll around in the bowl until, due to the friction, it finally settles down at the bottom.

Now we don't have a bowl in this apparatus, but we have something analogous to it. It's a minute trap that holds an ion within the vacuum by virtue of the electric forces it exerts. The trap (much the size of the eye of a sewing needle) consists of a stainless steel ring and two cap electrodes....A radio-frequency voltage is applied across the electrodes to create the electrical field of force capable of holding the ion.

If you were to shine sunlight on this barium ion—I don't think the experiment has been done!—it should look bluish-green, and so it is quite correct to think of a barium ion as a little blue-green dot.

What the ringmaster and the trainers of this atomic circus do is dispatch a very weak beam of barium atoms into the superb vacuum they have set up as their work space. (Unauthorized atoms, keep out!)

Another weak beam (of electrons) crosses the barium beam inside the trap, and collisions occur. Some collisions chip a little charge (one electron) away from a fraction of the passing barium atoms. Those few atoms are then no

 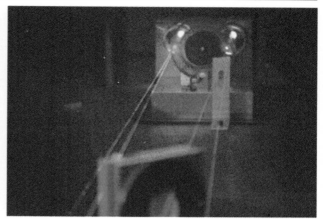

A glimpse of the atomic circus. The two spectacular laser beams are matched to the energy-ladder distances of the atom that will perform. Optical, electronic, and vacuum apparatus of various kinds may be made out. It is hard to catch an entire circus and its trainers in one picture.

longer electrically neutral, but are charged so that now they can be held by the electrical trap. In this incomplete state atoms are called ions; more broadly they can be regarded simply as certain special rather high-energy states of the original barium atom.

Some chipped barium atoms—ions—are thus trapped, held as though in a tiny frictionless bowl. These can gradually wriggle out of the trap by their random motion. But the experimenters have directed the two laser beams to cross the trap as well. They act there to provide some friction, to reduce the random motion of the trapped barium ions through many collisions with the powerful streams of subtly-tuned laser light. In effect, the laser beam is heated by the ion, and the ion is correspondingly "cooled"—that is, its random motion is greatly slowed.

The trapped incomplete atoms repeatedly jump up and down among three nearby energy levels of the quantum ladder, as either red or blue-green laser light is taken up or given out. The faster atoms leak out of the trap; slowly moving ones are counted by noting the amount of light they emit. The light signal increases as more charged atoms are created, one after another, inside the confining trap, then decreases step by step as they leave.

At some time the light signal indicates that only one single barium ion is left in the trap. That individual in the trap is excited over and over again by the laser beam. It emits its own blue-green light in random directions. Whatever light it happens to send toward the glass porthole can leave the vacuum to enter a focused camera waiting outside.

But the atom appears as a single tiny colorful dot!

What could you expect? Atomic architecture simply cannot be caught by any light that the eye can see.

It is the deep knowledge of its energy-level diagram, its quantum ladder, that enables such utter control of an atom. The experimenters can count off ordinary atoms one by one, then gently hold a single trembling specimen in place for hours, to pose for a picture in its own characteristic color during a million identical quantum jumps.

NO PICTURE OF AN ATOM AT ALL, NO VISUAL representation, is as useful as its set of internal energy levels, its quantum ladder, as we have said. Atoms are simply too small and too strange to be visualized in ordinary space. But the set of energy levels, along with the properties of each level, fully defines and marks the behavior of atoms, so that with that guide we get what pictures we can, and we can manipulate, shadow, arrange, measure atoms in many ways. Its quantum ladder is about the best representation we have for each atom.

Its own specific set of quantum levels is appropriate to any enduring structure, and it is not restricted to atoms. It is used by chemists

This is the first photograph in color of a lone atom, taken with ordinary film in an ordinary camera (with an extra-closeup lens). Exactly at the picture's center is the blue-green dot of the atom. It is an ionic state of barium, localized in space, alone and almost still, viewed by reflected light, one individual atom in natural color. (Some of the metal structure of the trap shows by stray light, red, white, and blue-green.)

QUANTUM LADDERS

THE ENERGY-LEVEL DIAGRAM—THE quantum ladder—is by no means restricted to atoms. It is useful very widely. Take as an example the oxygen atom. (Hydrogen is so simple that in using it a few possibilities are missed.)

The first few energy levels of oxygen are pretty high up; the lines are in the ultraviolet. Then as the energy probes higher and higher, the atom begins one by one to shed its electrons, and other ladders appear, one for each of half a dozen ions, as the incomplete atoms are called.

But inside the oxygen atom is its nucleus. If a sample of oxygen is exposed to energy input far more energetic than ultraviolet light, the levels of the nucleus will eventually show up. They form a whole new ladder, whose rungs are spaced roughly a million times farther apart than the atomic levels. Those levels describe the structure of the nucleus.

Nor is that the limit. Go fifty or a hundred times higher still in collisional energy, and you begin to see the levels that mark the internal structure of those particles whose mutual arrangements determine the nuclear levels themselves. Beyond that are new transient particles, made on the spot but not present initially in any ordinary matter. They have their own diagrams of internal energy.

Look the other way, toward more gentle probing. Oxygen atoms within the air of the room are always bound into pairs. That paired structure must have levels with a good-sized gap around the energies of the visible colors, for the air is transparent. If there were many level spacings within the ordinary color range, the matching light would be absorbed, and air would have a color. Down in the infrared there is another set of levels, tens of times below the visible, a range where oxygen is sure to interact. The air is not at all transparent in that spectral band.

Quantum ladders are everywhere.

who probe giant molecules with delicate radio techniques or in the deep infrared; and it is used again by those who send the hammer beams of big accelerators in to smash subatomic particles so as to reveal their structure. That very structure is expressed through their quantum levels.

The applications of the energy diagram are universal. The scheme has proved long-lasting. The idea goes back seventy years, and its use is unchanged today. Of course, the structures we study today are different. We still study atoms and their arrays, but we go far within the atom as well. The theories we have of what holds structures together improve. We find more and more new subatomic particles, and now we even know new subatomic forces.

But the description in terms of quantum levels remains the same. I believe it is a permanent part of our knowledge. I believe it will last as long as North and South America will be entered on sailors' maps, or as a full description of the solar system will include the many moons of Jupiter.

Ours is a quantum world.

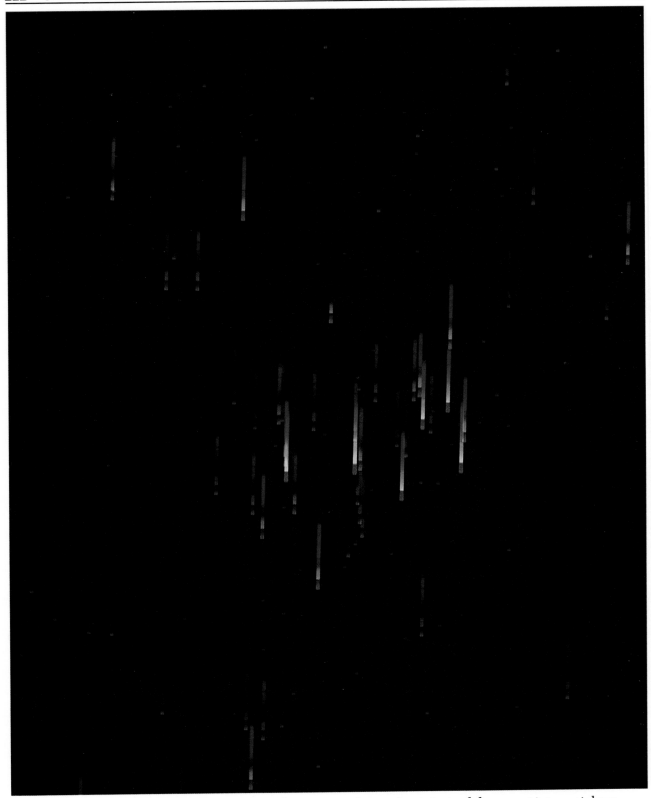

The Pleiades in color. Each bright star is imaged not as a single spot but is instead drawn out into a rainbow stripe, its spectrum.

6

ASSURANCE AND DOUBT

IN THE SKY OF A WINTER'S NIGHT THERE IS A tiny cluster of stars that many know and call the Pleiades. That tight group of stars has attracted the curious watcher for a very long time. In Augsburg in 1603, a few years before any telescope had looked at the sky, Johann Bayer and his talented engraver published a wonderfully illustrated book, the first of all printed star maps of the whole sky. It shows us just what the unaided eye could see in the heavens, mapped constellation by constellation. Naturally the Pleiades are present in a neatly engraved pattern of half a dozen closely set stars.

For a long time people liked to argue about how many stars could be seen in that cluster. The keenest eyes in the darkest skies could count ten or a dozen. (Look back in Chapter 1 at the stars of the Pleiades drawn by Maestlin in 1579.) But for more than three centuries hundreds of fine telescopes have scanned the night sky. Our knowledge of that sky is enormously richer.

Compare, for example, one section of a modern photographic presentation of the whole sky area by area, to see what that survey telescope could make out in the Pleiades. The same

bright stars are there in the same pattern; the eye is confirmed. But along with them there are large faint patches of light and many hundreds of much fainter stars that the eye can never see, clustered around and among the few bright ones.

That touches upon the change that has occurred in our perception of the starry sphere, now that newer instruments arm the astronomer with prodigiously extended vision.

Examine a still stranger photo of the same Pleiades in color. With a little attention anyone can make out that the underlying pattern of bright stars is the same. But now no star appears as a single dot or spot; instead the image of each one has been drawn out into a rainbow stripe, its spectrum.

The spectral photograph of the Pleiades was made with a modern telescopic camera at the Kitt Peak National Observatory in Arizona. A two-foot glass wedge placed in front of the aperture spreads the light of each star some distance across the photographic emulsion according to color, making a spectral record of all the stars bright enough to show.

For more than a century that kind of spectral study has harvested a windfall of knowledge

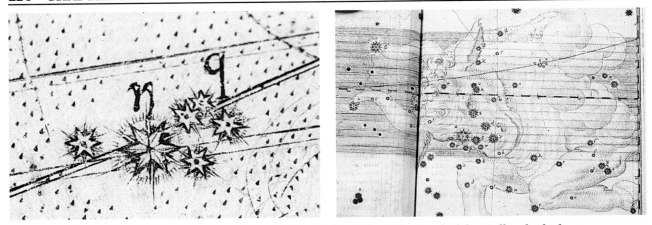

The Constellation Taurus in Bayer's star atlas of 1603. To the right of the head of the Bull is the little star group called the Pleiades, also shown enlarged.

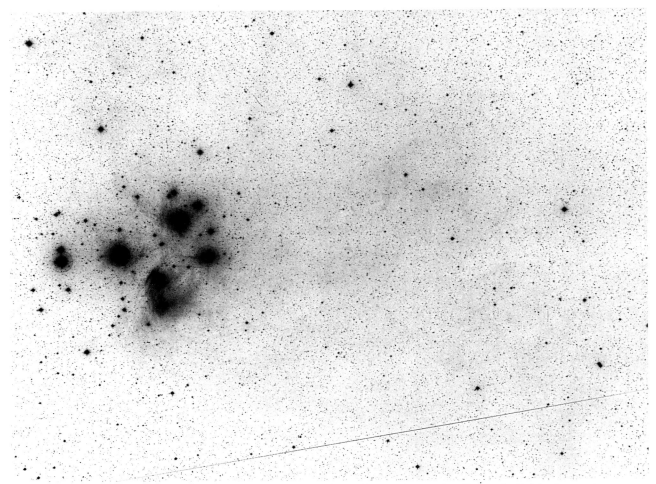

The Pleiades in a modern atlas of the whole sky, one made not by eye but with a fast wide-field telescopic camera four feet in diameter, using long time exposures on sensitive emulsions. The cluster is very rich in faint stars.

The Pleiades ride high in the winter sky, outshone that night by the passing moon. Here the camera reports the sky much as we see it.

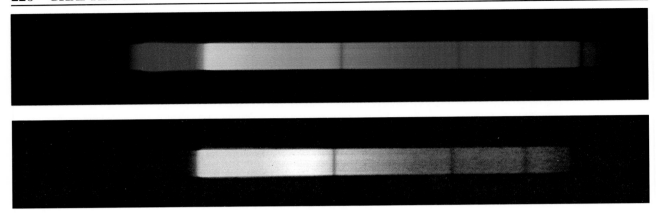

about the universe, insight we never expected to have. Once the physicists in the laboratory had understood how glowing gas makes light, we could read the recipe of the faraway stars, their chemical composition, from the details of those stripes of color. A big ground and polished wedge of glass, the equivalent of a prism set in front of the telescope, is as good as a teaspoonful of the Pleiades fetched magically to the laboratory across all the light-years.

A closer look is required to see just what the astronomers are looking for, just where the information is. We chose a typical star near the edge of the photographic plate and enlarged its spectrum. Now we see quite plainly what the astronomer might look for on the plate under a magnifying lens. The small spectrum extends from the red, color after color along the entire rainbow gamut to the other end, the violet. The features the astronomer seeks are many dark lines that cross the stripe. Those are called the lines of the star spectrum. They bear the information about chemical composition.

Long ago it was recognized that those line patterns, the sequence of positions of the lines along the color band, are a kind of fingerprint for a chemical element. Many elements possess a conspicuous set of such lines. Once you can identify a grouping of lines, there can be no doubt. Atoms of that element must be present in the luminous gas that is the source of the light. Now stars are certainly not great

The wedge of glass being shifted to the front end of this telescope is simply a circular prism. It spreads out each point of light—each star—into a colored stripe, a spectrum. The drawing may clarify the shape of the big glass element.

The spectrum of one star of the Pleiades, shown first in color and then in black and white. This spectrum has been enlarged to show the detail the astronomer looks for on the plate by direct magnification. Spectra in black and white are much less attractive than those in color. But notice that full information about the spectrum of our example is there still. The band of color still extends from red to blue as it did, even though the hues are not to be seen. The spectral lines remain visible. Measuring the position of those lines, to record where they lie along the color sequence between the limits at red and violet, is all the physicist needs to establish what he wants to know. We will use many professional spectral photos in black and white. They bear the full information of the spectrum. Color photography is still a luxury for the working astronomer, and often the black and white emulsions hold more detail.

balls of conveniently pure substances. They are natural mixtures. The spectroscopist must count on untangling many superimposed patterns of known atomic lines. The analysis is not easy. But it promised so great a marvel as knowledge of the inner nature of a distant star, and people therefore worked hard at the sorting. Let us follow the work a little way.

AT THE UNIVERSITY OF MICHIGAN, WE LOOK over the shoulder of a contemporary expert, Nancy Houk. She is an astronomer who is devoting years to systematic study and classification of stars by their spectra. The instrument used in Arizona to make the spectra we have been examining and another like it in Chile supply the data for her ambitious project. She classifies the spectra by noting the line pattern each has, and sorting them into meaningful categories according to the details of the pattern.

> HOUK: My experience has been built up over the years. By now I've looked at over 100,000 individual star spectra; this practice has made classification almost automatic. The eye is really quite marvelous in how consistently it can detect small differences in the strength of the photographed lines. One person can judge in a much more consistent way than a team can; that's why I am doing all the actual classification myself.

Astronomer Houk at Ann Arbor examining spectral plates under magnification.

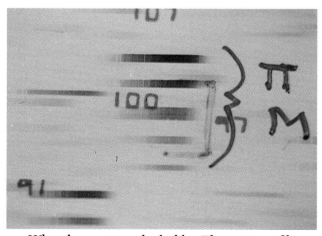

What the spectrum looks like. The pattern of lines is the basis of classification. Every star spectrum fits somewhere in the long sequence, arranged according to what lines are strong or weak, narrow or broad, in spectra of differing intensities.

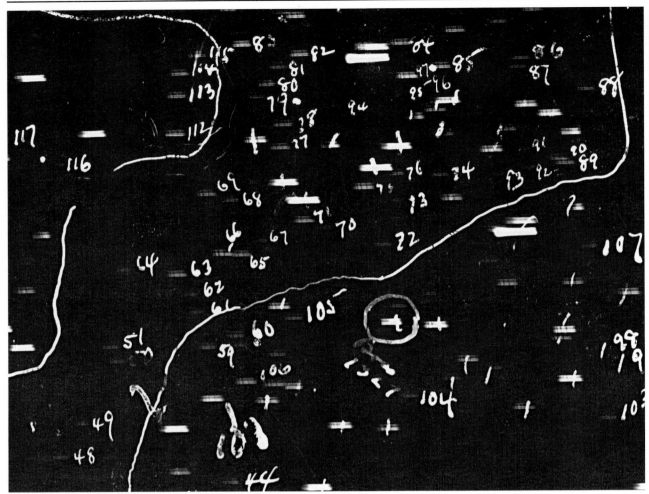

A spectral plate of a part of the southern sky exposed in 1895 from the Harvard mountain station in Peru, annotated and classified years later by Annie J. Cannon in her own hand.

The process I go through when I classify a star for the first time is by now mainly unconscious. I look to see whether there are many lines or just a few, whether they are narrow or broad, and whether there are any wide bands across the spectrum. Once I look at the spectrum I know approximately the temperature and brightness of the star from the details of the lines. Then I set a standard star spectrum alongside the one I am classifying, and compare the two to check it out.

We are more than halfway now. By the time we are finished we will have photographed the entire sky, from the South Pole to the North Pole, and I will have classified nearly a quarter of a million stars. Annie Cannon had by the 1920s

classified those stars into about forty-six groups, but with modern techniques I can sort them into several hundred categories. That will take a while; our whole project may be finished in about twenty-five years.

FOR A GENERATION THE HARVARD COLLEGE Observatory was the world center of stellar spectroscopy, begun as a long-range project before the turn of the century. Edward Pickering, who organized the project, Annie Cannon, who became its most celebrated specialist, and a generation of workers built up a treasury of spectral images from the stars (all those rainbow images are recorded in austere black and white). Nowadays the archives at Harvard

Astronomer Cannon at work on a spectral plate at the Harvard Observatory, using skylight reflected up through the plate from a mirror.

shelve more than half a million sky photographs of all kinds.

It was plain that here was an ocean of information; these recorded spectra of the stars are the only means to gain knowledge of their substance. When the photos were first made that application was not wholly foreseen; it came as the fulfillment of a long-held trust that reliable data would someday become of great value.

Harvard's record of a quarter-million star spectra was called the Henry Draper Catalogue, after the New York astronomer who first carried out photographic spectroscopy of the stars. The lengthy study was largely supported in his memory by Mrs. Draper, and grew over the years under the long hard work of a large staff, mostly women recruited for the decades of effort. Among them the most renowned was Annie Jump Cannon. Dr. Cannon became so expert that reputation ascribed to her the awesome ability to recall at sight just which plate bore every one of the quarter-million spectra she had arranged! Her nine-volume catalogue was completed in 1924. (The twenty-first century will use its successor, now being prepared by Nancy Houk.)

This unique library of star data is still in active use. The astronomical librarians who know and manage its riches today readily found for me one of the plates Annie Cannon had worked on. It resembled a plate of Houk's modern survey. Cannon had inked it with a spiderweb of numbers to identify the stars she had classified. The spectral patches are clear but very small.

It was these myriad small spectra she knew so well how to classify. She and her colleagues placed them after a while in an order into which they naturally seemed to fall: there was only a small difference in line pattern between any pair of consecutive members, but across the entire series the difference was large. What that order was due to, what it depended on, was not fully known to the astronomers of that time. There were speculations but no workable theory.

The spectra were not all sheer mystery in Cannon's day. The lines were evident, and their patterns could be identified as the lines of particular elements, the chemical elements well-known on earth and later in the spectrum of the sun. Not all stars seemed to contain the same list of elements; there were stars rich in hydrogen lines and other stars rich in lines of iron and other metals, for example.

Making allowance for the brightness of a star, before the twenties the Harvard spectra also could be seen to indicate the temperature of a

Dr. Cecilia Payne, brand-new Ph.D., seated at extreme right before a plate-study stand at the Harvard Observatory. The women of the research staff posed for a group photo in 1925. Antonia Maury is seated at left front, and Annie Cannon with her magnifier is third from left; they were then the best-known staff members. Payne-Gaposchkin later wrote of them: "the two women... were to be the great contributors to the birth of astrophysics...Miss Cannon had a genius for classification...Miss Maury had a passion for understanding...."

star's surface through its color. Red stars, say, had spectral bands that faded out before the blue end was reached. Such stars were red-hot like the iron cooling on the blacksmith's anvil, there was good reason to believe; stars with spectra that were bright all the way into the violet (red light was hardly recorded at all by the old emulsions) were white-hot.

A whole gamut of color could be laid down that went smoothly with the complicated line patterns. But why? The full classification included much more detail, and paid more attention to subtler distinctions than this simplification suggests, but it remained basically empirical, and almost entirely qualitative. In so long a list there were plenty of nonconforming stars as well, thousands of them, not all understood to this day. This ancient Book of Spectra was written in a language not yet readable, though here and there meaning could be puzzled out.

A Recipe for the Stars

A BRILLIANT AND CULTIVATED YOUNG English graduate student came to Harvard Observatory in the early 1920s, when the spectral behavior of the atom was beginning to become clear. Cecilia Payne arrived with that new quantum understanding. She knew how spectra worked, and she felt that she had both the zeal and the ability to puzzle out the unread hieroglyphs of the spectra of the stars.

She knew that they might contain the recipe for the cosmos.

Fifty-odd years afterwards Cecilia Payne-Gaposchkin, as a distinguished emeritus professor at Harvard, recalled her student days in an intense and artful autobiography:

In September 1919 I entered Newnham College, Cambridge. The atmosphere was euphoric. The "War to end war" was over. Few among us young people doubted that the Millennium was upon us....The intellectual atmosphere was equally heady....Physics was at the parting of the ways. The classical branches of the subject were indeed enough to fill the student's time if he pursued them in their lapidary symmetry and elegance through mechanics, electromagnetic theory and thermodynamics, but radioactivity was in the air. It dominated the Cavendish Laboratory, for Rutherford was beginning his attack on the atomic nucleus. ...The Bohr atom was introduced to us by Bohr himself...the New Physics was gaining momentum....The lure of the Cavendish Laboratory was irresistible.

Such was the scientific panorama as I saw it at the end of 1919. I was standing before the door through which I was soon to enter that world for myself. Suddenly, dramatically, it swung open....The result was a complete transformation of my world picture....I was dedicated to physical science, forever.

Harvard College observatory about 1920. Payne's new brick building is at the left.

The advanced course in physics began with Rutherford's lectures. I was the only woman student who attended them....

My college life drew to a close. Nearly all my friends were looking for teaching positions, for teaching at a girls' school was virtually the only kind of work to which we could look forward....My taste of the world of scientists had unfitted me for such a calling.

It was almost impossible for a woman, however talented, to follow such a research career in Britain. She was determined to be a research scientist. The New World offered more; she met and was invited by the young new director of the Harvard College Observatory to come to the other Cambridge as their first postgraduate student.

If Harvard Observatory in Cambridge, Massachusetts, its spectral library of the stars still unread, was the right place to come, Cambridge University in England was about the best possible place to come from. It had been about ten years since the theorist Niels Bohr and the experimenter Ernest Rutherford, working in close association, though not jointly, had built the first quantum theory of the atom, which lay behind any attempt to understand the spectra and substance of the stars.

Cecilia was armed with that theory. She had heard both Bohr and Rutherford lecture in person at Cambridge. The theory was actively growing in the hands of many people. She was an admirable student of the quantum theory, and she felt that she could apply it to the light from the stars. That work had been begun, but "only in the most general terms."

With a few key ideas—an understanding of the physics of glowing gas, the needed quantum ladder for many atoms, and the necessary identification, not of one but of two energy levels that went with every spectral line—she intended to decipher the recipe for the stars out of the shelves overflowing with Harvard plates.

Cecilia Payne sailed with high hopes for Boston from Southampton in the fall of 1923.

When I arrived in Cambridge, Massachusetts, I had crossed a gulf wider than the Atlantic Ocean. I had left the world of dreams and stepped into reality. Abstract study was a thing of the past; now I was moving among the stars....

The day after I arrived in the United States, I was installed at a desk in the "Brick Building," then the newest structure in the Observatory grounds....It had been built to house the great collection of photographic plates....Here also were the

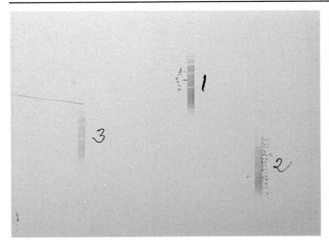

The spectroscopist's craft. The large bright spectrum (number 2) exposed many years earlier is one Payne marked to use as a standard set of lines.

Cecilia Payne in 1924, Pickering Astronomical Fellow at the Harvard Observatory, in front of Leverett House.

offices of those who worked with the plates.

When I first looked at the plates from which the Henry Draper classifications had been made, I was amazed. The spectra looked like tiny parallel smears....It seemed impossible that anyone could see enough in those tiny smears to classify the spectra. Sometimes, indeed, I would find one of Miss Cannon's numbers in a spot where I could see nothing but a faint blur.

HERE SHE WAS AT TWENTY-THREE, READY TO analyze the distant stars.

She herself exposed few or no plates at the telescope. Plates were already here in plenty. But she had in any case to become a remarkable craftsman in a trade she more or less invented. There is a lot of technical work before you can tease out quantities for a recipe from those little patches of gray on the photographic plates.

The plate librarian at the observatory brought out one of the very plates young Cecilia Payne had measured. These are spectra of single stars, spectra much larger than the little images made when the plate records the spectrum of every star in a field all at once. More is to be asked of these larger and more detailed spectra. We have Payne's notebook by which we could verify that the writing on the plate was indeed hers.

After all, this plate is standing in for a teaspoonful of star. It samples light alone, but it can lead us to substance. Payne studied the old spectrum (taken long before in Harvard's field station on a mountaintop in Peru) in quite a new way. She had two tasks. First of all she wanted to judge whether a line was weak or strong. She did that by eye, simply comparing it to the other lines on many photographs. Second, she had to be sure that she understood the origin of every line she used. That is to say, she had to place it in its proper interval on some atomic quantum ladder. Otherwise she could not use the line for her analysis of the recipe of the stars. A line that could not be placed properly in the quantum ladder of an atom could indicate only that the atom was present, because the line had been observed in the laboratory spectrum of that atom, but not the amount of it.

It was clear that some quantitative method must be devised for expressing the intensities of spectral lines, and I set up a crude system of eye estimates. Next came...the arduous task of estimating their intensities on hundreds of spectra....

There followed months, almost a year, as I remember, of utter bewilderment. Often I was in a state of exhaustion and despair, working all day and late into the night....

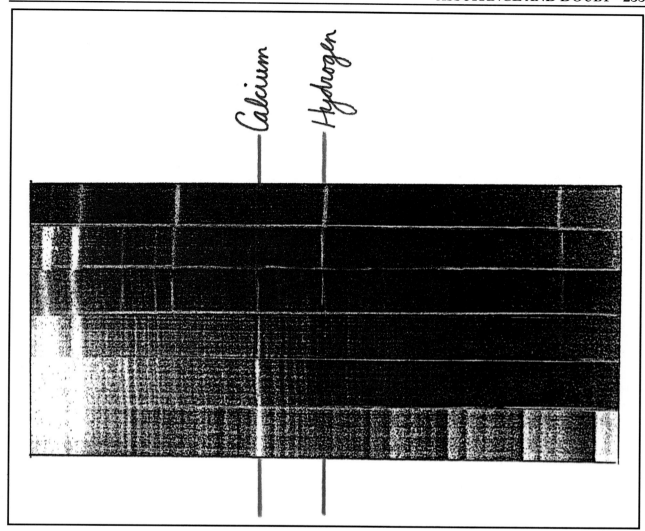

Calcium   Hydrogen

Finally some light dawned in the darkness.

WE CAN FOLLOW THE LOGIC OF PAYNE'S argument without her powerful mathematics. One example is a violet line she reported as coming from the atoms of the element calcium. We all recognize calcium as a major constituent of bone and shell, and an essential nutrient. Calcium lines were well-known in the spectrum of the sun and of most stars.

But about this particular line of calcium she knew a decisive additional fact. She knew that it occurred only when an atom of calcium resting in its lowest possible energy state picked up just the right energy to jump to the very first rung of the calcium quantum ladder.

Violet lines in stars hot and cold. Six spectra of different stars are arranged by the temperature measured for each star, the coolest at the bottom, the hottest at the top. Two lines in the violet are marked. The one on the left is the line of calcium Payne studied; plainly it fades away in hotter stars. Near it is the violet line of hydrogen. It appears only in the hotter stars. Why the difference?

She searched out that line, not any line of calcium but that particular line of calcium, in star after star. She arranged the sequence of spectra to run from the coolest stars, barely red-hot, on to the hottest, blue-white stars. The series showed an unexpected but absolutely unmistakable regularity. In the cool stars, that line was strong. It changed in a smooth way from star to star. In the hot stars, that same line was weak.

Earlier it had been thought that such a result must mean there was no calcium in the hot stars; they were different stars, and lacked the ingredient calcium.

But that is not at all the case. We can understand it just as Cecilia Payne worked it out on the basis of quantum theory. If a star is very cool, most of the calcium atoms must be down there in their lowest state. It is the energy they gain and lose in their random collisions with light and with other atoms that can excite any atom. Atoms jump to higher states and down again, or may jump down only to go up, over and over. Thus atoms are found at any one time spread among many of their energy states. Now in the cool stars—provided there is enough light to see them at all—you find a very strong line of calcium.

It is easy to see why. Most of the calcium atoms in the cool gas are not much excited; they are in the lowest state, the one state that can give rise to the line under study. But if the star in question is very hot, the calcium atoms there can pick up a good deal of energy in many different collisions. So in hot gas the atoms are found with enough energy to spread them among many, many energy levels. At any one time every atom is in some single energy state, but there are many possible higher-energy states, each of which will be occupied by some of the atoms. Then the number of calcium atoms left in the lowest state, the unique state from which they must start if this particular line is to be present, must be much reduced. The particular violet line of calcium—not all the lines of calcium—is then understandably strong in cool stars, and systematically weaker among hotter and hotter stars.

Cecilia Payne could even treat the question quantitatively, though without mathematics it is tedious to try to follow her in detail. She could estimate from the statistical theory what the temperature would have to be in a gas that showed a weak violet calcium line, and check if that was consistent with her calculations of gas temperature from the color of the star, a method that used no data from individual spectral lines at all. The whole picture was thoroughly consistent. The calcium line was not exceptional. Many other lines of other atoms behaved the same way.

Then she looked hard at a contrasting case. By chance that example also used a nearby

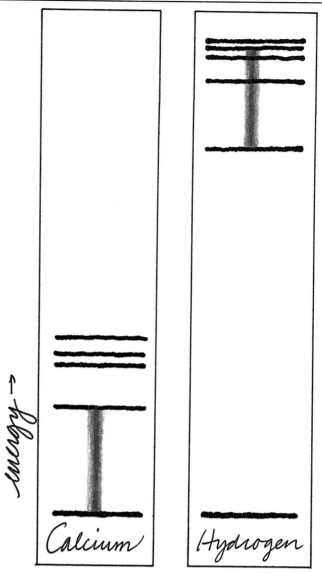

blue-violet line, but this time the line belongs to the element hydrogen. Hydrogen is, of course, part of water, hence abundant in our bodies, in all living matter, and in the sea. (Hydrogen atoms are single in the hot gas of these stars, where the atomic cluster H_2, common on earth, is unstable.) Some hydrogen lines are present in nearly all the stars, but they are conspicuous only in a few.

If one sets out again the cool red stars and follows the temperature sequence step by step on to the hottest blue-white stars, this violet line of hydrogen also changes markedly in intensity. But its behavior is just opposite to the behavior of the calcium violet line. The hydrogen violet line is relatively weak in the red stars, but very strong in white-hot ones. Why?

The hydrogen line that Payne looked at is a line that cannot arise in a jump from the lowest state of the hydrogen atom up to its second state. On the contrary, for this line the atomic jump begins in hydrogen level number 2, and ends well above that in level 5. But in a cool star not many atoms of hydrogen at any one time can be as high as energy state number 2. Indeed, they are very rarely found there; nearly all of them must remain in the lowest level, seldom excited across the first energy gap up to state 2. In fact, that step is remarkably large; the light that it corresponds to is deep in the far ultra violet (much beyond the mercury ultra violet line we looked at in the previous chapter).

Energy-level diagrams for the calcium and the hydrogen atoms. For each atom the violet lines of the stellar spectra Payne analyzed are shown. The calcium violet line begins at the ground state. In cool stars plenty of calcium atoms are resting in that state; the line is strong. In hot stars, few atoms are there, and the line is weak. For hydrogen just the reverse is true, since that violet line can arise only from atoms that are well above the ground state. So the hydrogen violet line is strong only in hot stars.

There is simply not enough energy in the weak collisions of the slow-moving atoms in a cool gas to excite hydrogen atoms to the second state.

Only as the temperature rises do more and more hydrogen atoms temporarily pick up enough energy, either from light or from collision with some other atom, to push them up to the second hydrogen state. Yet only those atoms that are in the second level can possibly contribute to the hydrogen line Payne found.

Therefore, it makes very good sense to say: yes, this particular line of hydrogen must be weak in cool red stars and strong in hot blue-white ones. And that is the case.

Payne's confidence grew as she found agreements like that for a few dozen lines of different atoms. You begin to trust a theory that works so well. "Two years of estimation, plotting, calculation and the work I had planned was done. I had determined a stellar temperature scale and had measured the astrophysical abundance of the chemical elements."

CECILIA PAYNE'S WORK IN THOSE YEARS HAD a remarkable outcome, expressed nicely in her published thesis of 1925. People have called it, and I think they are right, "the best thesis ever written in astronomy."

First of all, she made it plain that the composition of most of the stars is broadly the same. The gross differences that we see from star to star in spectral lines were not at all due to different recipes for the different stars, but only to

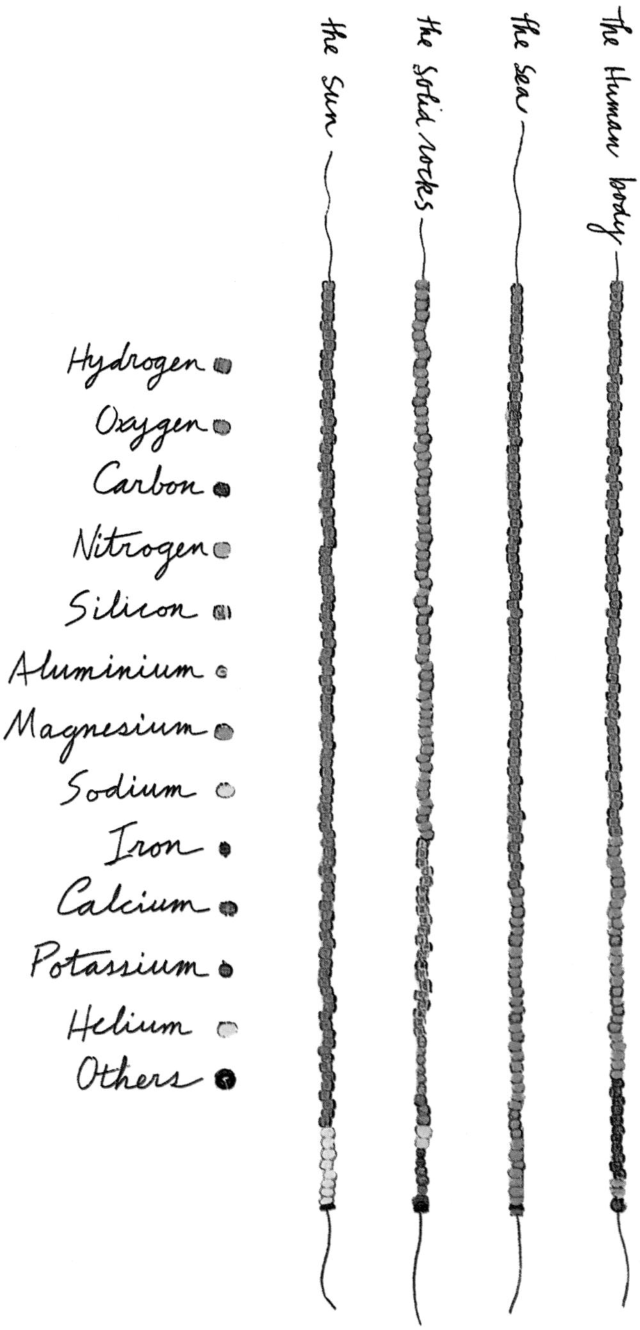

Hydrogen
Oxygen
Carbon
Nitrogen
Silicon
Aluminium
Magnesium
Sodium
Iron
Calcium
Potassium
Helium
Others

the Sun · the solid rocks · the sea · The Human body

The ingredients of the world: four current labels. The atomic ingredients in four interesting structures are given in a simple form. The numbers given tell how many atoms of each kind there are among a hundred atoms total, within the human body, the sea, the solid rocks of earth, and the sun. (Only those elements that contribute at least one atom out of a hundred are listed.)

different temperatures of the same common gaseous mixture of elements. Star stuff is in general quite uniform.

But she found something more, a conclusion so startling that she herself and her senior advisers could not quite believe it. The hydrogen violet line was as strong as the violet line of calcium, even though it could come only from those few atoms of hydrogen that happened to be on the energy-rich second rung of the quantum ladder. But if only an exceptional hydrogen atom could form that line, and yet that hydrogen line appeared as strong as a line of calcium to which almost every calcium atom present could contribute, there was only one admissible conclusion. The hot gas of the star must contain many more atoms of hydrogen than it did of calcium. In the same way she could find telltale lines for many atoms that would give her the abundances of the elements relative to one another.

Her calculations led to an incredible result. In a typical star there are a million atoms of hydrogen for every one of calcium, and far more hydrogen atoms than all the rest of the elements put together. She had put a label on the stars that listed their ingredients.

In her thesis she retained the ingrained caution of theorists applying a new theory for the first time. Payne wrote: "The outstanding discrepancies between the astrophysical and terrestrial abundances are displayed for hydrogen and helium. The enormous abundance derived for these elements in the stellar atmosphere is almost certainly not real." Most of the other elements were present in proportions that were similar enough to those of earth. Of course, there were scores of elements she did not find at all. When these were common elements, though, the reason was usually clear: they had no lines in visible light that were strong lines and well-assigned to levels.

Now we believe she was right in the first place; her caution, no doubt urged on her by her advisers as well, turned out to be misplaced though entirely understandable. That enormous hydrogen abundance is real. The universe we see is hydrogen-rich. In fact, everything that we can see glowing in any of our instruments, apart from our own rocky little earth and still lesser bodies, seems to be made primarily of hydrogen. Today a commonplace of the textbooks, in 1925 that was a revolutionary recipe for the universe.

After I had finished this first essay in astrophysics, I went to see Eddington. In a burst of youthful enthusiasm, I told him I

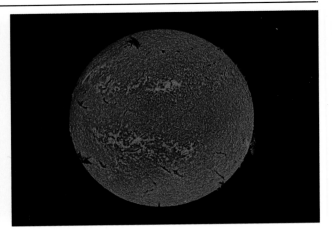

Hydrogen is everywhere; the spectra disclose its presence in glowing gases around the bright young stars, in our sun, in the giant gassy planets, and in the far galaxies.

believed that there was far more hydrogen *in* the stars than any other atom. "You don't mean *in* the stars, you mean *on* the stars," was his comment. In this case, indeed, I was in the right, and in later years he was to recognize it too.

It was hard to believe that 99 percent of all atoms in the universe should be hydrogen and helium, familiar but far from so predominant here on earth.

Scientists were prudent to doubt it then for a while. The first version of the quantum theory that it was based upon was seriously incomplete. But within five years that theory matured enormously, and the techniques for investigating spectra photographically also improved. At the end of those five years the most expert critics were satisfied, and themselves loudly proclaimed that the universe was made mostly of hydrogen.

The fact is more important even than it sounds. It is not only that it was unexpected, that it dealt something of a blow to our long-held Copernican expectations to learn that the matter around us every day was not a fair sample. The result itself was full of meaning.

Consider the two atoms that are the most abundant. More than ninety atoms out of a hundred are hydrogen, and helium makes up most of the rest. Those are not just two random elements. Hydrogen is the lightest and the simplest atom that can stably exist out of about one hundred species of stable atoms, and helium is the second simplest and lightest!

The matter of the universe is therefore fundamentally simple. Its atoms are young, pristine, somehow little evolved. That's what we have to infer from Payne's dazzling result. And every theory of the origins of the universe has to come to grips with that fact.

Confirmation

ALL THE SPECTRA OF THE ASTRONOMERS until the 1940s were obtained with visible light, or perhaps just beyond the margins of the visible color range. But after World War II there was a new technology of microwave radio, largely from its utility for radar. Interest

A PRIMER OF CELESTIAL SPECTROSCOPY; WHAT A STAR SPECTRUM CAN TELL US

1. CERTAIN LINES, ESPECIALLY IN COMbination, show that an element—atom or ion (incomplete atom)—is present in the glowing gas. If more than one substance is present, more than one set of lines appear.

2. Each single line tells of an energy transition made by atoms from one level to another.

3. The intensity of the lines tells how often an event (an energy transition) occurs. That rate depends both upon how many atoms of the substance are present and in which of their levels the atoms are found.

4. The temperature of the gas in the star can be read by the lines of the atom that occur, for in hotter stars atoms excited to higher levels will be more common. The result should in detail confirm the cruder judgment of overall color, in the sense that bluish stars are hotter than red ones.

5. The contrast of lines against background (brighter or darker) tells whether or not the light passes through a thinner and cooler gas after it was emitted. All lines of normal stars are seen dark against a background that comes from the denser hotter material below the outer atmosphere.

6. The position of recognizable lines (shifted redward or blueward compared to their laboratory position) tells the speed toward or away from the observer that the emitter is moving.

And that isn't all.

The simple copper horn, pointed roughly at the center of the Milky Way, with which the microwave line of atomic hydrogen from the cold dark gas between the stars was discovered.

grew even outside of the astronomical world among physicists and engineers, eager to follow a few early hints that a new view of the universe would open up in those new channels, new extensions of the visual sense.

In the lab it had been shown that the lowest state of a lone hydrogen atom carried a delicate complication, foreseen by the theory. That lowest level was not single, but split by a very small energy gap into two levels. No quantum ladder we could draw on ordinary scale shows that split; that energy gap needs to be magnified something like a millionfold before it can be legibly drawn on the quantum ladder of hydrogen we used for the visible spectrum.

The light that would be emitted by that tiny quantum jump is not visible light at all, but radiation far beyond the red end of the spectrum, indeed far beyond what is usually called the infrared. It is in the microwave region, the same band of radiation produced in microwave ovens, quite like the band that carries the UHF channels of television.

The laboratory physicists had measured just where that line would be found in the microwave region. So small an energy of excitation could be present frequently even within a very cold gas, as long as there were plenty of hydrogen atoms present. With that idea, a physicist at Harvard and his graduate student, a veteran of wartime electronics, thought of trying a new sort of spectroscopic astronomy. They would look for that radio line from hydrogen atoms in the cold interstellar spaces, gas far from stars or nebulae, out there in the darkness.

Harold Ewen and Edward Purcell let us in on a personal recollection of their celebrated discovery of 1951.

PURCELL:...it's right here, I think this is the window...our horn came out of. We just took the glass out and put the throat of the horn in, and looked out south past the sun. The astrophysicists...had guessed that there was hydrogen away out there that was much colder than anyplace on earth. The radio power at that special frequency amounted to only one watt spread over the entire earth. By any measure we were looking for a very weak signal.

EWEN: I was concerned that I might be dealing downstream somewhere with a negative Ph.D. thesis. If you don't detect something, then you must carefully state at what level you are capable or incapable of detecting it, or the negative result is of little interest. Ed's comment was: "So it's a couple of years of your life...but it's certainly worth it. And if you do detect it, you will be in *Life* magazine." He was right.

PURCELL: Well, as I remember it was morning. He'd been up all night and I'd been at home in bed. He phoned, "I think I have a thesis," and I came dashing over.

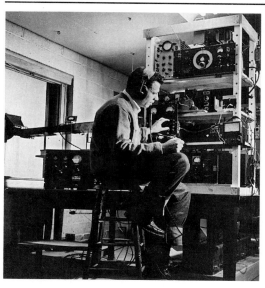

Doc Ewen amidst his electronics, vintage 1951.

EWEN: It was Easter weekend. The first time the rig was turned on I tuned through the spectrum as you might turn a knob. At the end of the first scan, the signal was on its way up. There was a pen recorder marking off the signal on a long roll of paper.

PURCELL: We rolled about twenty feet of chart out here in this corridor and we got down and sighted along it, which was very exciting.

EWEN: There it was, the very curve we had looked for, just about what we were thinking it might be like if the hydrogen line was a bit above the continuum. It was crude, but it was there.

PURCELL: It wasn't too long before we were pretty sure. We didn't have quite the ironclad proof to publish yet, but we knew it was real.

EWEN: I recall discussions that Ed and I had early on, selecting a topic for my thesis. Then…deciding on what sort of an instrument to build. After that, actually building it, trying it out. Then, when the switch was turned on and you tune through the line, there it is the first time, no problem. It's just the way you designed it. It's just the way you thought about it. A chill goes up your back and you say, "I've got it." You never, ever forget the excitement of doing something like that. It's beautiful."

That was one more confirmation of the universal hydrogen: much of the cold empty space between the stars of the galaxy is in fact filled with free single hydrogen atoms. They are few and far between; that space is much closer to a vacuum than anything we can make in the lab, but there is space in God's plenty, so that a tenth as much hydrogen as we find in all the stars of the galaxy is out there in cold dilute form, as lonely atoms. Once more the element hydrogen dominates.

PURE HYDROGEN GAS IS A BULK INDUSTRIAL commodity here on earth. Yet the familiar terrestrial form of hydrogen was not known in the interstellar spaces. At modest pressure and temperature hydrogen atoms cannot remain single; each clings to its partner, a twin. The two atoms bound together are called a hydrogen molecule, H_2. If that is broken apart into H atoms, each of the pair of atoms, of course, shows the familiar atomic spectrum, with many visible, ultraviolet, and infrared lines, and the famous microwave line as well. But as cold H_2 the stable twinned structure has no important spectral lines in the visible, the infrared, or the radio regions.

We knew that in the lab the two clinging hydrogen atoms did show strong diagnostic lines in the ultraviolet, but those lines are in a color range where radiation cannot penetrate the earth's atmosphere. You cannot see that form of hydrogen in space by looking out from

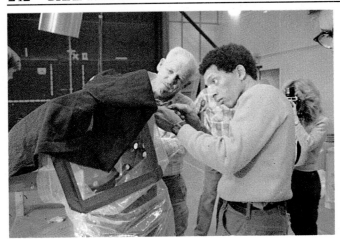

George Carruthers at White Sands Missile Range in 1986.

the ground. Somehow the experimenter must circumvent that natural atmospheric filter. Here then is a thoroughly modern form of astronomy: send your ultraviolet spectroscope up in a rocket!

That is just how George Carruthers, young physicist at the U.S. Naval Research Lab, became the first to find molecular hydrogen, the dominant constituent in the cloudier and denser portions of interstellar space. He launched the new instrument he designed for the purpose in a sounding rocket from the White Sands Missile Range in New Mexico in 1970.

We found George Carruthers getting ready to rocket up the ultraviolet spectrograph once again from White Sands in 1986. (This time his aim was Halley's Comet—and he found atomic hydrogen there, though molecular hydrogen has not been seen.) He recalled the old days easily:

CARRUTHERS: We had to develop a very sensitive instrument in order to obtain the necessary data in a sounding rocket flight that lasts only about five minutes, while the rocket soars 120 miles high, well above the absorbing air. In the early 1960s the technology was not very advanced. We had many failures before we had one success. Not only did the rocket have to go above the atmosphere, but also it needed an attitude control system that pointed

with sufficient accuracy at the stars we chose.

The unique features of our instrument were that it was especially designed to be sensitive in the very short wavelength range of the ultraviolet where molecular hydrogen has its lines. Secondly, we used an electronically intensified camera system that was much more sensitive than conventional photography. We were able to look at faint, obscured stars, important because molecular hydrogen is expected in rather dusty places among the stars.

We were always excited to get data from any sounding rocket flight, since we knew it was something unique that no one had ever seen before. In the case of the molecular hydrogen we knew at once that the rocket experiment was a success, but it took several weeks of looking at the data before we really could say we had seen molecular hydrogen in the spectrum.

Not long afterwards other groups launched a satellite into orbit where it could circle longer than any sounding rocket could spend above the air, to observe not for minutes but for years. A big space probe like that is well beyond the reach of any single investigator, even one ambitious enough to take on the responsibility of a sounding rocket with its launch team. That probe, named Copernicus, carried a sophisticated ultraviolet spectroscope that has since

A prudent check on contents without ever seeing them.

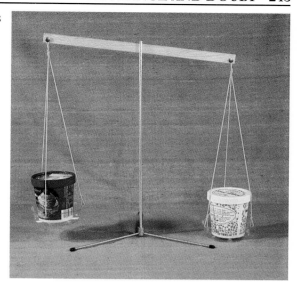

given us much richer detail about molecular hydrogen in space. But first of all to find molecular hydrogen within the clouds of space was Carruthers with his five-minute sounding rocket.

This last form of pure hydrogen to be found in the cosmos was the very kind most commonplace here on earth. There seems to be about as much molecular hydrogen as there is atomic hydrogen in the galaxy as a whole; but it was much more securely hidden from our view. Molecular hydrogen was the final entry in the long list of unanimous reports from every band of radiation that together single out hydrogen atoms as the chief ingredient of the cosmic recipe.

Doubt

OVER THE DECADES RESULT AFTER RESULT extended Payne's much-confirmed analysis. She had demonstrated that the ordinary stars, which are most of what we can see, are predominantly made of hydrogen and helium.

Whenever we looked with new means into the heavens we saw hydrogen everywhere: in the nebulae, those warmer clouds of hydrogen and helium that glow so extensively—if so faintly—in red and green within the space between the stars; in the sun's cold giant planets, spheres made mostly of hydrogen and helium; in the swift particles of the cosmic rays, mainly hydrogen and helium. The radio astronomers reported mostly hydrogen in the cold gases of space. Even the rocket-borne detectors brought back the same news: gaseous hydrogen in the denser interstellar clouds, in a form common on earth, but earlier unseen in the skies. Wherever we look, whichever radiation we examine, the report is the same: a universe dominated by the simplest, lightest, smallest atom, hydrogen. The astronomers looked and reckoned for fifty years; the better they got, the surer was the conclusion. Everything we can see out there is rich in hydrogen.

That is a result of first importance. We need to rest a lot of confidence in it, for we can never understand the rise and fate of our universe until we know what it is made of. Is there any way to check up even roughly? There is one way, a commonplace, earthbound procedure, familiar to prudent shoppers for centuries. I brought two similar containers home from the corner grocery. Each bore the same words on the label: vanilla ice cream, one pint. They looked the same size; they claimed to be the same material. Can we check the recipe in any fashion?

Indeed, we have one check that does not even ask us to open a carton. All we need do is to weigh the two samples. It is easy to balance one against the other, and perhaps only a small surprise to find that of two pints of vanilla, each the same size, one is a good deal heavier than the other. We have found something out about the contents without looking inside the carton. The method can assess the invisible. Inside the lighter carton some lighter ingredient is more abundant than it is in the other carton. That's about all we can say.

We cannot be very knowing about it; we cannot, for example, conclude from this result alone what that light ingredient might be. Those who know something about ice cream will recognize a likely candidate, especially when I add the hint that the heavier carton was also the more expensive carton. But we could not tell from this external but powerful method of disclosing the invisible just what it was.

To go beyond what any instrument of vision in any color can tell, we need some procedure for weighing distant substance. The scales seem very far from astronomy; but what the astronomers have to do is to find a method to weigh plenty of stars and gas in space.

THERE ARE INDEED MANY COLLECTIONS OF stars, samples that just might be weighed in bulk. They are the galaxies; they lie far, far beyond any of the stars we see in the night sky. Stars are grouped in galaxies much as leaves are grouped on trees. If you lived in a big tree, and looked out to see the other trees of the wood, many leaves of your own tree would be conspicuous across the foreground of your view. Peering between them here and there, you could make out many other trees a little way off. The nearer trees might be visible in detail; on some close trees even individual leaves could be distinguished. As you looked farther out into the woods, you would see more distant trees as green-textured forms, their leaves no longer separately visible.

That analogy is very much like our real situation. Every one of the stars we see at night with unaided eye is a foreground star in our own galaxy; call it a leaf on our own tree. The band of dim light that we name the Milky Way is an edgewise view of a large disk of stars, a disk that stretches out so far away that the stars merge together in our view, though they are within our own galaxy. The unaided eye can see only one or two external galaxies—some other trees of the woods—in the whole sky. But the big telescopes see a whole wooded landscape, with trees and clumps of trees as far as vision goes.

Ours is a typical good-sized galaxy, with a couple of hundred billion stars. The sun, our star, is therefore minute on the scale of a galaxy; even the whole solar system is much smaller than is a leaf compared to a tree. Those stars are gravitationally bound, so that they all revolve in orbits mutually attracted to the more central bulk of stars. Looking out in the clear spaces past the foreground stars of our galaxy, astronomers with big instruments have studied the galaxies large and small. They resemble our own galaxy more or less as trees broadly resemble each other, though no two are quite the same. The galaxies are spaced rather like trees in open parkland, nearest neighbors usually a dozen or so galaxy-diameters apart.

There are some 10,000 or even 20,000 galaxies close enough so that we can study them in some detail with the big telescopes, though hardly a hundred are close enough to pick out bright individual stars. We know that the stars of the galaxies resemble our own, and we know as well from the spectra of the galaxies, using the light of many stars that merge into one spectrum, that the other galaxies, too, consist of stars and gas made mainly of hydrogen and helium.

We can even identify a great host of galaxies that lie so far away that their tiny fuzzy images give us only a little overall information. There must be about a billion galaxies like that grouped and clustered out to the present distance limits of our best instruments; no one has seen all of those images, but we estimate a count by sampling as the pollsters do.

IN THE HEAVENS WEIGHING IS DONE BY THE analysis of motions under gravity; no one imagines we might literally put a galaxy or even a planet on a pair of scales. Here the earth pulls upon each pan of the balance; any difference in that gravitational pull tilts the beam. It is the same force, gravity, that acts to control the orbits of moon around earth, of the planets and their moons, of sun and planets, and of the many pairs of mutually bound double stars we see within our galaxy. All that has been checked with high confidence; we can predict orbital motions in detail. The reliability of the almanac is proverbial.

We believe the galaxies, too, are places where myriads of orbiting stars are bound by their mutual gravity. There is no detailed check, but rather general consistency with many tests, to be sure less searching than the precision forecasts of the solar system. The weighing of a galaxy is a process that follows the same principles as the weighing of earth or sun, a procedure familiar to Isaac Newton. The application to such great distant structures as the galaxies raises a special challenge, but it can be done, and it is not hard to follow.

The unexpected reward was an astonishing result.

It was Galileo who first began to grasp that every object fell under gravity in the same way, no matter what material or how heavy it was. We have watched the feather and the hammer fall together on the airless moon, and we know that a little grapefruit-sized satellite and the big spacelabs both circle in low earth orbit all the way around in the same hour and a half.

Then it was Newton who worked out the details of orbits under gravity. Newton's result for circular orbits, directly tested both by th

LOOKING OUT INTO THE WOODS: A GALAXY PRIMER

HERE WE PRESENT EVIDENCE FOR what galaxies are, results that follow without any calculation or instrumentation beyond looking at and thinking about a few sky photographs.

What are galaxies?

The first pair of pictures in color were made by a very expert Los Angeles amateur astronomer with his own telescope. They show a pinwheellike patch of blue light in the direction of the constellation Coma Berenices. The scattered stars in the image are the leaves of our own tree, the few stars that we see in this direction.

The two pictures show a change rare in such an object. In the top photo there is a very bright star, not seen at all a year later in the lower picture. New stars that rise and fade are nothing new. The Crab Nebula was like that in 1054. But we see them like that by unaided eye only every few centuries (one in 1987!). We do see a dozen or two every year in the telescopes, always in or very near a little patch on the sky like this one. What can we conclude? This is indeed another tree like our own, with many leaves—many stars—and rarely one becomes a supernova, revealing the starry nature of an external patch so far away that like a distant tree in the woods its stars merge together to our telescopes. A supernova so far away is telescopic; nearby it would dominate the night sky. Since each patch is a whole galaxy, we ought to find a supernova every century or so in any one galaxy. But since we see hundreds and thousands of galaxies in the telescope, we should expect to find a supernova quite often in one or another outside galaxy. And we do.

Galaxies are distant groups of many stars, more or less like the Milky Way.

...h and without a supernova.

The Whirlpool with its spiral arms is a circular disc, seen face-on; it has a small companion.

M81 is somewhat elliptical.

The Andromeda Galaxy, M31, a narrower ellipse.

Galaxy NGC4565, seen edgewise.

What shapes are they?

A dinner plate on the table looks somewhat elliptical as you look at it from where you sit at table. Look straight down; it is circular. Across the table the plate is a still narrower ellipse. Edgewise it seems a thin linear structure.

Four photos of different galaxies show the same forms. We conclude that these are all circular, but viewed under more or less tilt. We live inside the Milky Way galaxy as though we lived in the material of a dish; it looks like a band of tight-clustered stars all the way around our sky. Looking at right angles we see far fewer stars, through the thin direction of the plate, not its diameter. Most galaxies are seen in the two parts of the sky where you look out the thin way. Put yourself in thought within the galaxy of the first picture, NGC 4595, and the situation will be plain.

Many galaxies like our own are thin circular disks of stars around a central bulge. (There are other types of galaxies as well.)

How big are they?

Some trees are small and some large in any woodland. Distant trees always have small images, and show less detail, even if they are good-sized. But trees might be small or large, near or far, and it would not be easy to be sure if the small tree you saw was a large one far away, or a smaller one not so far.

The four images are reproduced at the right relative size from a series of plates all to one scale made by one instrument. The size of image and the detail may give some idea of the size and distance of each galaxy, but more information would be needed to gain any sure results on size or distance. That is a big task for students of galaxies.

Galaxies come in a range of sizes and can be seen over a wide range of distances.

On the names of galaxies:

There are 10,000 or more galaxies close enough or large enough to show up in some detail. Most of these are listed in a compilation called the *New General Catalogue*, begun in the last century. They bear serial numbers with an NGC ahead of the number. The showiest few dozen belong to a list made by Charles Messier in the 1770s; they bear an M followed by a number up to 100 or so. They are the most often illustrated. A very few have pet names, like the Whirlpool.

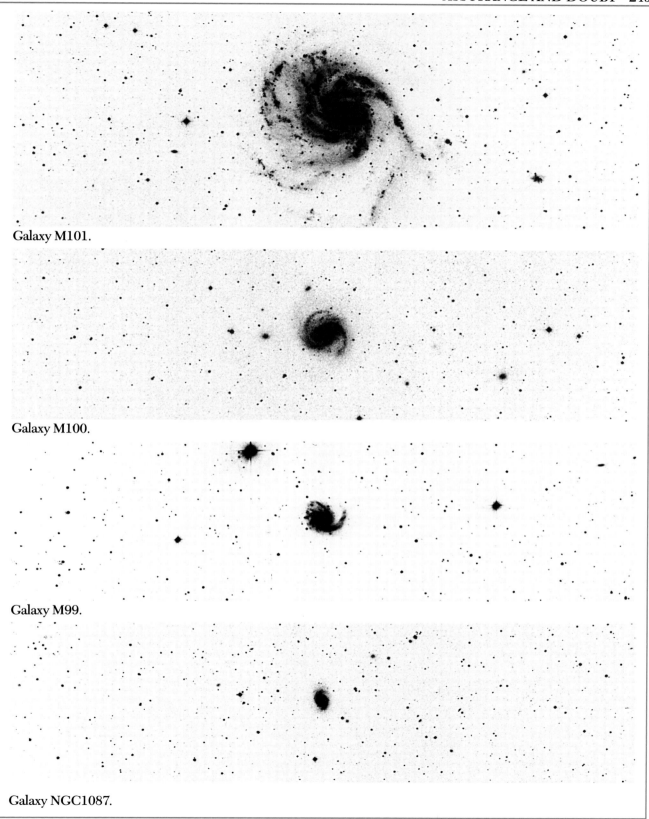

Galaxy M101.

Galaxy M100.

Galaxy M99.

Galaxy NGC1087.

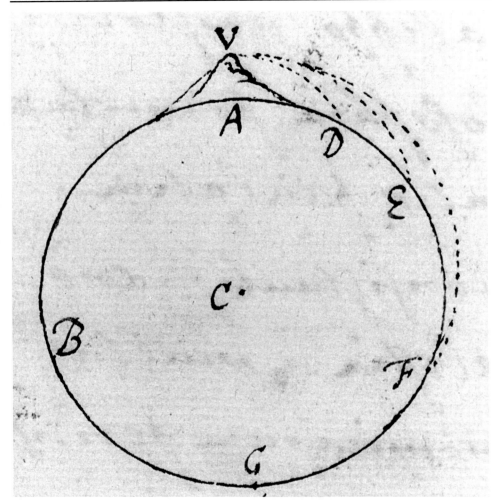

Newton's own sketch explaining the orbit of an earth satellite, prepared by him for the engraver who drew up the published version reproduced in Chapter 1.

satellites of the solar system and by the artificial satellites he had himself imagined three centuries ago as a thought experiment, is simple and general.

The speed of any object that moves in a circle around a gravitating body depends only on the size of that circle, for any particular gravitating body, say the earth, or Jupiter, or the sun. But for different attracting centers, the speed in orbit is the greater the greater is the weight of the gravitating center and the bigger the central pull. The orbiting body in effect weighs the matter that is pulling it in; what controls its motion is the matter, all the matter, that lies inside its orbit circle, and that is the weight it reports. The orbiter is only a witness; its orbit depends not at all upon what it is itself but only upon the attracting matter within the circle. By the study of the orbits, the astronomers could

give fair values for the mass of earth, Jupiter, Saturn, and the sun by the middle of the eighteenth century.

For the solar system the sun is central. The nearest planet to the sun is Mercury; it rounds its small orbit in only three months. Mercury's circle is a little under half the size of earth's; its orbital time is a quarter of a year, the measure of its higher speed in orbit. Saturn takes thirty years to traverse its big circle, but that journey is only ten times as far around as our annual jaunt on earth. The farther away a planet is from the sun, the slower the speed with which the planet moves in its circle. The pull of the sun is diluted by the greater distance, and a slower orbital motion is the requisite response to reduced gravitational pull.

Can we weigh the sun, then? Indeed, here is the procedure. For each planetary circle, a

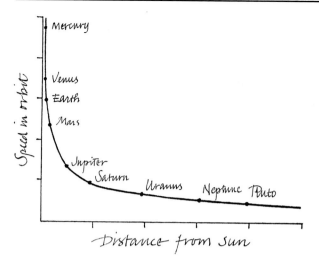

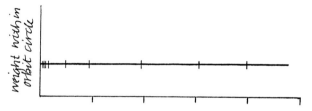

Newton weighs the solar system. Two graphs sum up what Newton's theory prescribes for the solar system. The first curve is the rotation curve observed for the planets, from Mercury out to distant Pluto. On the vertical scale is plotted the speed at which each moves, and on the horizontal one the distance of each from the sun. Earth is the third planet. We see that Mercury and Venus move far more quickly than the outer planets. By applying Newton's method to each of the observed planet speeds, we can calculate what is shown on the second graph. There each planet declares that the same total weight—mass—is held within its orbit circle, as the flat line shows. Essentially the whole weight of the solar system is that of the highly visible sun, central to all the orbits. You get what you see.

simple formula of Newton gives us the weight of all the matter enclosed by each orbit in turn. The result is the same for all of them. Point by point, planet by planet, all the orbital tests concur; the only important weight enclosed within the orbits of the planets is that of the sun itself. The plot of the weight or mass that is within a given circle around the sun rises only within the sun, and remains flat and constant outside of the visible disk. Essentially all the mass in the solar system is that of the central sun. Of course the planets contribute something to that mass, but their total is quite negligible compared to the dominant sun. The weight of Earth is equally to be found from the circles of its satellites, natural and artificial. They concur; our planet has a mass about one part in 300,000 of that of the sun. The sun is huge, too; the volume of its fiery sphere is a million times the volume of earth.

NOW THE WAY IS CLEAR TO WEIGH THE galaxies.

The galaxies are simply stars (along with gas and dust) in motion. If we choose a star or a group of stars that circle a galaxy, all we need

do is watch that bright spot move slowly from place to place around the center of the galaxy, and begin to build up the rotation curve for the galaxy. Once that is plotted out for a good sample of circling star orbits, we can find the total mass within from gravitation, as we weighed the solar system. That offers a check that the galaxy as a whole has the mass we impute to it from its star composition and size, or simply by comparing the stars we see there to the sun and stars of our own galaxy, where stars have been individually weighed in many cases by the use of circling starry companions we can see in the telescopes. (So far we know only one planetary system, ours.)

A simple but daunting barrier is in the way. Motion is not easy to find on the prodigious scale of galaxies. Think of the solar system. The planet Saturn circles in stately orbit around the sun in thirty years. Watch for a month and anyone can see it move against the star background even without magnification. (It takes some effort and thought to eliminate the effects of our own moving planetary platform).

But galaxies are built to a scale enormously larger than the solar system. Our sun, for instance, orbits in its grand circle about the distant gravitating center of the Milky Way in a matter of 200 million years. At such geologically slow rates of circling, it would take 10,000 years for some feature in a nearby galaxy to change its position relative to the galaxy center by the smallest amount the telescope can detect. The farther galaxies would be slower in proportion. Distance conceals motion. Astronomers are proverbially patient, but a wait that long is ridiculous!

There is happily a quicker way. Some hint of it comes from the motorist: you can judge the speed of a car by watching the apparent motions of the landscape. Or you can look at the speedometer. The point is that we can think of motion in two ways: a changing sequence of positions over time, or a judgment of rate made in a tiny interval. It turns out that the physics of signal propagation (for light and sound alike) offers a way to judge speed directly, not by waiting for position changes to take place over considerable time.

Here is an experiment to show the wonderful effect. We begin with a small electronic tonemaker, a coffee can packed with battery, electronics, and a loudspeaker. The speaker emits a steady, loud, rather pure musical note as long as the switch is on.

The athlete vigorously swung the electronics-packed can around and around his head on a six-foot cord, handle at arm's length. Just as a check, we attached a microphone to the coffee can, so that we might pick up the tone emitted while the source was in motion. The steady pitch of the tonemaker was unaffected by the swinging as heard by a microphone moving along with it. But when we changed the placement of our microphone to listen to the tone from a fixed point as the source kept moving in its ten-foot circle, there was a dramatic difference. We were listening from a point well outside the circle; the microphone picked up the moving source from a position more or less edgewise to the circle of motion.

The motion of the tonemaker relative to the microphone altered what we heard. The pitch was distinctly higher whenever the speaker approached the microphone, lower as it receded.

The source was naturally louder or quieter as it came closer to or farther from the recording microphone. To make it easier to hear the effect we wanted, a change not in loudness but in pitch, we selectively sampled the sound in time. We just cut the recorded tape to listen only to the moments that the tonemaker was passing the very ends of its orbit, farthest to the left and the right of the microphone. At those two times it was moving, first straight toward and then straight away from the mike.

When it came toward the mike, the pitch was higher. When it went away from the mike, the pitch was lower. In between there was a

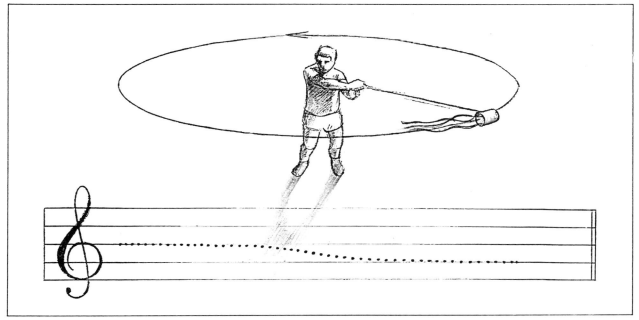

Doppler shift. The athlete orbits a can of electronics around and around his head on a long rope. The electronic tonemaker produces a steady note. But when a stationary microphone reports what it hears, the tone is changing. The diagram shows the motion of the source of tone; the musical scale gives us the changed notes we heard at the two ends of the orbit. The pitch was audibly down a little at the end where the source was going away from the microphone, but up when it was coming toward it.

A video astronomer at work among the electronics consoles and monitors: Vera Rubin during an observing run in the control room of the four-meter Mayall telescope, the largest on Kitt Peak.

transition; when the loudspeaker was moving neither toward the mike nor away from it, but only across the path that the sound took to reach the mike, the effect was minimal.

Think back to the drummer whose taps helped us map by sound in Chapter 3. Suppose we heard a regular series of taps. If the drummer is not moving, the time intervals we hear between distant taps would be the same as those the drummer used. But if the drummer advanced toward us as she tapped, successive beats would have smaller distances to travel. Then the sound of each tap would be delayed less than the preceding one: the drumbeats would follow each other to the ear more closely than they had been produced.

The magnitude of the effect here was reasonable. For this speed, about twenty-five miles an hour, the note we heard shifted by one semitone on the usual musical scale. The faster the speed, the greater the effect. Plainly pitch change can be used to measure the speed of motion of a source of tone.

The same effect is present in light, too. Color is here the analogue to pitch. A moving light source of well-defined spectral color shifts slightly redward going away from the observer, and blueward coming toward one. The radar speed monitor of the highway police sends out a fixed radar signal, and compares the quality, color if you like, of the return echo shifted by the motion of the echoing automobile. The astronomers are blessed by the same phenomenon. They can measure the speed of many light sources by measuring the shift in position of any spectral line coming from the source, which will be shifted redward by a receding motion, blueward by approach.

The effect was first predicted by Christian Doppler long ago, and it has been verified many times for light, radio, and sound. By direct measure of speed it circumvents the problem of hopelessly slow changes in galactic position over a human lifetime. Orbital speeds are not at all slow among the galaxies; it is only that the distances are formidable.

With this phenomenon, using light, not sound, we will measure the speed of galaxy rotation by a meticulous study of the spectrum. From the speeds we can form a rotation curve, then a mass curve, and thus weigh all the material visible or not that is in a galaxy.

The extension of the sense of color by the spectroscope made colored light an amazing means of remote study. Already we have seen how to find the temperature and the chemical composition of distant glowing gas. Now we see that light can be used to measure the mo-

Many domes, many telescopes at Kitt Peak National Observatory. The unusual peak near the middle of the skyline is in Papago Indian lore the center of the universe.

tions of distant light sources as well, as long as they emit some well-marked spectral lines. The spectroscope is as much an essential tool of astrophysics as the telescope itself.

Astronomer Vera Rubin of the Carnegie Institution of Washington has put these principles of physics to work to weigh galaxies. She looks at the numerous galaxies in the middle distance so as to have a wide choice of specimens, using one of the big telescopes both north and south of the equator. Let us visit this artist of the spectrograph to see just how she weighs a galaxy.

The largest telescope at Kitt Peak National Observatory on its Arizona mountaintop about forty miles west of Tucson has a mirror four meters (twelve feet) in diameter. A large telescope like that, an optical spectrograph able to spread light into a very wide spectrum, and new electronic image-amplifying devices to record the faint light in that spread-out spectrum have for a decade allowed close study of the rotation of galaxies.

The complex of instruments is rare and costly; astronomers eagerly apply for the

chance to use it. Available time is oversubscribed threefold. The astronomers whose applications have been successful are usually allocated use of this big instrument for only a few nights at a time, once or possibly twice during a year. Rubin and her colleague John Graham had three choice spring nights in 1986 in the dark of the moon, and they invited us to visit them at work.

They hoped to image a dozen different galaxies if the moonless nights were still and clear. An astronomer at a big telescope is a little like the captain of a small ship; there is a helmsman, the telescope operator, who actually handles the controls that point the telescope. There is an automatic guiding system to keep the pointing exact. There are electronics technicians who have made sure the detector is working. These days the telescope is guided not by looking through an eyepiece but by using a video monitor that carries the image

from the telescope in its open unheated dome to the comfortable control room.

Rubin spoke of the experience as she sat amidst the monitors, control panels, and electronics racks in the control room, her eyes on the video screen where the galaxy she had chosen appeared.

RUBIN: Telescope time is precious, so you must come with a very well planned program. You think you know what you are going to do every hour. Typically the exposures I make take two hours. It may take ten minutes to set up on the galaxy. You have to turn the telescope until it looks at the right spot. These are very faint galaxies; you must not make a mistake. No one wants to spend two hours looking at the wrong object. You work very hard in the hours when you would normally be sleeping, and you have to take care not to make simple mistakes.

You arrive on the mountain with a carefully worked out schedule, and then all kinds of things happen. It may be too windy to look in one direction, so you are permitted to use the telescope only in a direction you did not expect. You face hard choices. Perhaps there is one object you very much want to study. Should you do it the first night to make sure, even if the night sky is pretty unsteady, or gamble that the next night will be better? Most of your gambles are prompted by the weather, some by difficulties with the elaborate equipment you must use.

What happens when you work at larger and larger telescopes is that you see less and less of the sky. When you stand at a smaller telescope, the dome is open to the sky and you are right there to watch all night. It is in principle possible to come here for a three- or four-night observing run and never even see the telescope in the open dome. The detector is put on the day you arrive. You focus it. You get it all set up exactly as you want it. The rest of the time you could sit at the control console and never see the sky.

But if everything works well, it's an exhilarating experience. The sky is so very beautiful. Astronomy is a beautiful science. Most astronomers don't lose track of that; what we try to do during long exposures is get out and watch the sky with our own eyes. It's useful to make sure you know what the sky conditions are. But it's also a joy.

IN A FEW LONG INTENSE NIGHTS AT THE TELE-scope an astronomer collects data she may spend months to analyze. I visited Rubin at her research base at the Carnegie Institution, where she shared her exciting results of the last few years.

A computer terminal displays her work. She has plenty of photographs, but the most helpful images are on tape, easily and flexibly pre-

GALAXY ON THE SCALES

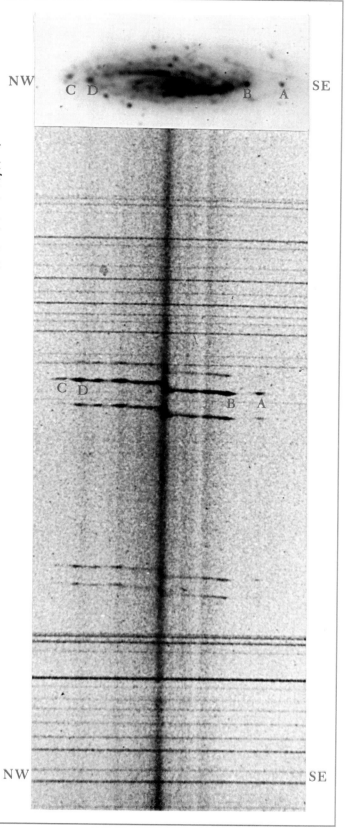

THE SAME GALAXY, NGC7541, IS NOW shown in negative, bright sky and dark star images. This galaxy is a couple of hundred million light-years away, a mid-distance object (only about a tenth of an inch across on the Sky Survey plate). The fuzzy patches are not single stars but more like whole constellations, each with many bright stars and glowing gas. It is those patches whose merged spectra are measured.

The slit of the spectrograph is placed along the galaxy to admit the light only from the unshaded area on the image. The spectrum is spread out very widely in the vertical direction, as the musical scale was drawn for the Doppler experiment. All that is seen here is a small region in the red. (The blue end would be a couple of feet above the picture.) Only the conspicuous broken lines belong to the spectrum of the galaxy, and the strong patches of light from the galaxy show up as knots in each spectral line. The strongest spectrum line is the red line of atomic hydrogen. It is seen displaced a little redward at one end of the galaxy, a little to the blueward at the other end, and it breaks from one place to the other rather sharply as you pass the center of the galaxy.

Other galaxy lines confirm the result. (The fainter unbroken lines do not come from the galaxy at all, but from the faint night glow of the earth's upper atmosphere.)

Just as for the tones we made in the meadow, the shifts of the red hydrogen line show that the galaxy is rotating, the patches of light in its right-hand end turning away from us, at the left end turning toward us. At the bright center the motion in rotation is mostly across the line of sight. The relative displacement of the spectral lines gives us the speed at which glowing matter at various positions in the galaxy is moving.

sented on the monochrome screen. I was impressed by the thick sheaf of photos she produced to show the galaxies she and her partner have weighed. I had expected a dozen examples; "Kent Ford and I have studied maybe a hundred now," she said. Rubin recommended that we look at "one of the nicest" specimens, by name NGC 7541, one of the galaxies open to close study by big instruments. It lies in the area of the sky we call the constellation Pisces, where, of course, the galaxy is far in the background, too small and too faint for any eye to see, a million times farther off than most bright stars of our sky.

She entered a few key strokes, and soon the right image showed up: a dark spindle-shaped pattern in lacy detail against a blue-green screen. There was a smaller companion galaxy on the screen beside it, spaced from it by about the size of the larger galaxy. We chose to look at the bigger one of this pair. "Here is the right image now, enlarged. I can also transform it into a reversed image bright against a black sky, a natural way to look at it. That choice shows the galaxy pretty much as we actually observe it."

A dozen or two bright spots stand out over the whole form of the galaxy. Those bright patches are not stars; individual stars are not to be seen. The patches are luminous gas clouds surrounding whole groups of stars, not unlike our own bright constellations—for example, Orion. The image was made by long exposure so the faint glow shows well.

Rubin created a line on the screen that crossed lengthwise down the axis of the galaxy spindle. That is where she set the narrow opening of the spectrograph when she was at the telescope. The light that forms the spectral line comes only from this path across the galaxy. At one end a few bright patches of the faint outer region contribute, then the brighter central portion, and at the other end the path crosses another bright patch or two, not quite so far out from the center.

The spectrum of the galaxy viewed along that path gives us the information we need to weigh the galaxy. All that is recorded is a short stretch in the red, a small part of the whole rainbow band, which on this scale would extend five or six feet from red end to blue end. That red region included the position of the strongest line being emitted by those glowing gas clouds, the famous deep-red line of atomic hydrogen.

As you look along the line you can make out the part of the galaxy that supplies the light at each place. At one end three bright clouds show up as bright knots in the line; the strong center is the bright nucleus, and near the other end there is a weak segment where the light came from an obscured part of the galaxy. We know that the hydrogen atoms in all those places are radiating the same line in the right intrinsic position, but the line has not recorded in the same position along the spectrum at all points. The line is broken by a jag. At one end the line is shifted redward a little; at the other end blueward. The effect is small; the eye would not notice any color change, but the spectrograph is more discriminating. The shift is only by a tenth of an inch out of a spectrum five feet long, one part in 500.

How can one end of the galaxy differ from another in the precise position along the spectrum of the hydrogen line emitted? The atoms at the two ends are certainly not strange new forms of hydrogen, but one and the same type of atom. But the two ends of the galaxy can very well be moving differently; moving sources of light appear with shifted color exactly as moving sources of sound appear with shifted pitch. This is the Doppler effect in light, known for a century and a half, and entirely analogous to the Doppler shift in sound. Motion toward the observer produces a higher pitch and a blue shift; motion away, a lowered pitch and a red shift. The faster the motion, the larger the shift in either sense. The opposite shifts, similar in amount at the two ends of the line, tell us that at one end of the galaxy the emitting clouds are coming toward us; at the other end, they are moving equally rapidly away.

Just as the hammer thrower pulled hard on the cord to keep the little electronic sound source whirling steadily in a circle about his head, so the gravitation of the bulk of the galaxy pulls the bright gas clouds around in a huge circle that orbits the galaxy. The part of the light in the spectral line that originates near the central portion of the galaxy comes from stars and gas that are moving neither toward us nor away as they go around the orbit circle, but moving only athwart the line of sight; there is no Doppler shift from rotation along that part of the orbit.

The galaxy is spinning around its center; all its stars and gas are moving, each in its own place and with its own speed. With the shifts that have been measured we can find out the speeds of rotation for a whole series of distances from the center of the galaxy.

There are data enough for some checking. The spectral position is carefully measured (under the microscope) all along the line. On one side relative to the center the cloud is moving toward us, the one on the other side is moving away. If the motion were truly circular, the motion away would have speed equal to the motion toward us. Results from the two sides are plotted together; they agree quite well, so the observed motion is close to a circular orbit. A weaker red line of a different atom, nitrogen, can be used instead; results agree with those from hydrogen. In a few cases one can find lines that come from star surfaces, not just from the big gas clouds; agreement again. Stars and gas move together as under gravity they must.

A graph can be made for each of many galaxies that plots the speed of motion observed in larger and larger circular orbits. That is the rotation curve of the galaxy.

Rotation curves for galaxies are fifty years old and more; they were valuable even when they

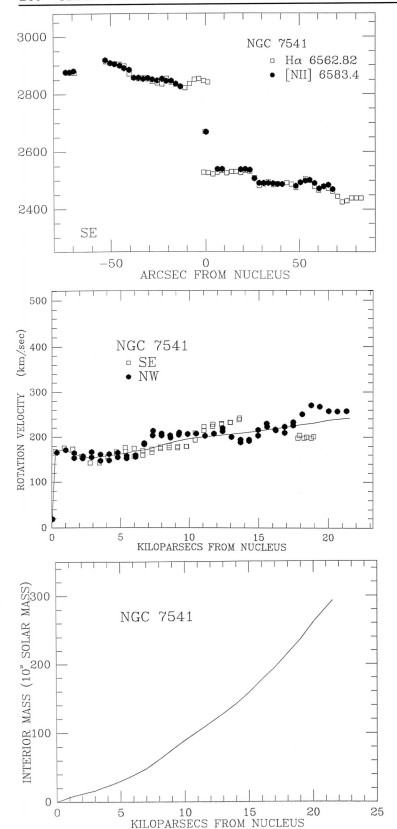

A galaxy weighed. Three graphs sufficed to weigh the remote galaxy, using the insights of Newton and Doppler with the instruments of today. In the first we see the rotation of the galaxy, the speeds plotted out from the observed spectrum for both the approaching and the receding ends. Since the circular orbit is symmetrical, the two ends were folded together; it is the same speed coming and going. The second graph shows how the speeds depend on the size of the orbit. You can check that the data from the two ends do fit well together.

The third graph is the payoff. For each orbital circle the Newton formula is applied exactly as though the star patches were circling planets. At each distance the measured speed declares the total weight held within each circular orbit. The galaxy has been weighed.

But the flat line we got for the solar system result is no guide to this galaxy at all. Here the weight grows and grows; the curve of mass goes steadily up and up as you look farther and farther out, even though the light of the galaxy is strongly concentrated in the center. You get much more than you see. Most of the galaxy is dark matter.

were rough and partial, because they confirmed nicely that galaxies are made of a gravitating host of orbiting stars. Here the old results of Newton are extrapolated to distances perhaps a million times longer than the reach of tested reliability here in the solar system. The possibility of error is certainly here, but so far we have not been misled.

The early galaxy rotation curves were quite satisfying to us solar-system residents. Like the solar system, the galaxies had very bright centers, not so condensed as a central sun, but visibly with the lion's share of light—and of stars. So the rotation curves behaved as they should: stars near the center of the galaxy move faster and faster as you look to larger circles, since the pull they feel is the larger the more light is included within the circle. That speed peaked just about at the edge of the bright galactic core (not far within the analogous place where the sun circles our own galaxy). Outside that place surely the orbital speed would go down, just as it does beyond the sun in the solar system, the starlight and hence the stars having run out.

Trouble was, the measurements stopped there, too; they couldn't then do a systematic job of looking out in the fainter regions. The occasional result that did come in from time to time did not show what everyone expected, but it could be set aside as an oddity.

Modern astronomy was developed by astronomers all of whom were familiar with the solar system. The central sun is dazzlingly bright, and the planets are mere specks of light. Almost all the mass is in the sun, and not even a part in a thousand more is to be found in planets, moons, and all the rest. The distribution of light is very similar to that of the mass or weight; what you see is what is there, at least roughly.

When you look at almost any galaxy, again the central portion is very bright and the outer parts faint. It was natural enough therefore to assume that the distribution of light in a galaxy is much like the distribution of mass, as certainly it is for the sun and the planets.

BUT WE WERE IN FOR A BIG SURPRISE.

In the galaxy NGC 7541 the orbital speeds rise very steeply as you look to points farther and farther away from the center of the galaxy. They still rise gently at larger distances, never to fall again, out to the limit of all the galaxy we see. The speeds rise as you look farther and farther beyond the last significant area of starlight. Astronomers have looked for faint stars in those darker fringes, and they just aren't there.

A rotation curve like that of galaxy NGC 7541 presents us with a quandary even at first sight. How can that distant orbital speed be kept up? Why is the pull of all those bright central stars not diluted by the big distances, down by a factor of a few at least?

The weighing procedure, done exactly as for the solar system, puts the case bluntly. The measured pull of the whole interior weight to which the distant orbiting clouds respond does not go down, but goes up, as you look to larger and larger orbits. There must be more stuff within to pull on the orbiting clouds; the mass encircled rises as you look to larger distances. There is almost no more starlight seen at those large distances from the center. Yet the amount of gravitating matter present beyond the edge of the central bright region is greater than everything that shines within.

That which has most of the weight does not shine, visibly or in any other wave band we know. We call it dark matter. All the spectral lines of hydrogen, all the matter we have found by its appearance in radio or X-ray or any other color whatever is only a small part of what is there. The stars and gas we see in the galaxy were weighed in the balance and found wanting. You get ten times as much as you see.

"When I started observing significant numbers of rotation curves of galaxies," Rubin recalled, "it became clear that none of the stars far out in the galaxies had a low speed. That was totally contrary to our expectation. It was not hard to accept because the observational evidence was so direct and strong. Nature has played a joke on the astronomers. We became astronomers to study the universe, and now we find we're studying only that 5 or 10 percent of it that is radiating!"

Orbiting patches bright enough to measure are seen only out to some limiting distance; we don't know how far out from the galaxy center the dark matter continues. Radio emissions of cold hydrogen gas way out there in a few galaxies suggest that the dark matter spreads even farther than the very farthest optical measurements. Dark matter is distributed in a big sphere around the galaxy center, and not in a disk, even in galaxies like our own where most of the starlight comes from a flattened central disk. But weighing a hundred galaxies has told us that most galaxies—our own probably included—consist principally of dark matter, by weight 90 or 95 percent of dark matter, invisibly distributed far beyond the starry image.

Rubin summed up:

We don't know at all what the dark stuff is. We don't know very well how much of it there might be. The more we observe, the more doubts are raised. Doubt is helpful. It wouldn't be as much fun to do science if every time you got an answer that was the end of it. Perhaps the strongest thing you can do as an observer is to find something that shows you need to learn more.

Astronomy has learned a lot since Cecilia Payne did her work. Although it took time for her rather indirect results on spectral lines to become accepted, once they were confirmed everyone could treat them as fact. There are many problems in astronomy at the present time that are even more indirect. The observational evidence is very complex; you have to handle and sift the data, go through lots of theoretical argument, before you come out with the answer.

But for the rotation curves that is not so

at all. You can look at the observation just as it comes off the telescope, and you know that what you are seeing is what is happening. Our observations of galaxy rotation are simple, direct, and clear. But the results we found are a true puzzle, for they tell us that there is something major we just don't know.

For fifty years news of the cosmic recipe was good. Every new channel we opened up brought back the same confirming message: yes, the material out there is mainly hydrogen. That was secure; we liked it, and it made a great deal of sense, for hydrogen is the simplest of atoms.

Vera Rubin and her associates worldwide with new techniques and crisp observational tests have changed that complacency utterly within one decade. Strangely, they didn't even show us that we had been wrong. It is still literally true that everything we see out there is made of hydrogen. But they added one overriding fact: by weight there is ten times as much matter out there as all we see!

How that could be no one yet understands. Is there something seriously wrong with the laws of motion or of gravitation, so that our inference about the weight is grossly in error? That does not seem likely, though it is certainly possible; we have made a wide extrapolation past tested results, and a few people are looking for an explanation that way.

Is there some way to package abundant hydrogen so that it will simply not glow, either by visible light or in any part of the spectrum that we can examine? It could not be present as a gas, or we would see it in X-rays or in ultraviolet or radio. Perhaps the galaxies are really swarming with a great many very small and invisibly faint orbiting dwarf stars made of hydrogen, each one about the size of the planet Jupiter. They might be missed. No one knows how such swarms could be formed, especially without a fraction of detectable bright stars among them. There would have to be a thousand times as many of those faint stars as of all the other stars we now know. Again, it is possible.

Perhaps there is some new kind of all-pervasive gas, particles never yet found on earth left over from some early stage of the universe, a kind of matter that cannot glow at all, but only pulls and weighs. That is possible, too, even rather popular among the physicists.

I am sure that most of the stars and the bright nebulae and the cold and dense gases in space are made of hydrogen. Cecilia Payne and the others showed us that. All that hydrogen is plain to be seen, if not in visible light, then in some other channel. But today we think that most of the matter out there extends dark and invisible beyond the visible margins of the galaxies; we cannot see it at all in any channel we now know.

If you ask me what the universe as a whole is made of, I must admit that as of now I remain in great doubt. I just do not know. But one thing I do know: we will try very hard to find out.

EPILOGUE

The task of science is both to extend the range of our experience and to reduce it to order, and this task presents various aspects inseparably connected with one another. Only by experience itself do we come to recognize those laws which grant us a comprehensive view of the diversity of phenomena. As our knowledge becomes wider, we must even be prepared therefore to expect alterations in the point of view best suited for the ordering of experience.
—*Niels Bohr,* Atomic Theory and the Description of Nature, *page 1 (1929; translated from the Danish)*

To reduce experience to order is a task older than science, the ancient goal of the philosophers. But to augment experience is the task that engages the artist, the artisan, the traveler, and indeed every child, who is the learner in us all. It was out of the slow fusion of the two that science appeared glowing in the crucible of history, and its composition is still visible in its affinities: science owes heavy debts to philosophy, art, craft, exploration, and childhood.

Popular accounts of the sciences have come more and more to emphasize the quest after order. We have here sought to redress the balance, and to place first the augmentation of experience. This cannot be naively done if science is to result, as Bohr implies. Neither every explorer nor every child is a scientist. But it cannot be neglected either; reciting the names of the planets and learning letter-perfect the great results, from DNA to the drift of the continents, is not science either, though it is admirable. It is worthwhile as well to assess the impact of discovery on our lives, even urgent in these days of a linked world that is nonetheless cleaved both north and south, east and west, but that too is not natural science.

Some sense of judgment, some degree of participation in science is a necessity if alienation is not to dominate the response of most people to this world of swift technical change. Some view of the experiences that underlie claims of order is needed if authoritative conclusions are to make sense. Evidently the experience cannot be complete; thoroughness defines the professional and not the general reader. It is the very richness of the experiences of science that limits the expertise of every working scientist to a rather narrow domain. What we have tried to present, to be sure only vicariously by word and image, is a varied set of specific experiences, each carried a little way toward the order it generates—or tests. Then a reader or a viewer may enjoy a taste of the tough tasks and the unforgettable pleasures of science, a treat not opened by abstract conclusions alone.

These six chapters, each of which comes out of a single hour of film made for television, offer a good many paths into the natural sciences. Here at the end of the text I will share the hopes of a pathmaker, not the scenes along the path, but rather why each path passed the way it did, to close with some comment on paths not taken.

The first chapter, "Looking," opens with an example of how far the images on the screen (or the words of a book) fall short of experience in this world of four dimensions and many channels. Then it develops the astonishing degree to which our inbuilt senses mirror the instruments of science. We infer the world around us every day from what we perceive exactly as the sciences do, with a share of their successes and their failures. It was one grand example, the telescope of Galileo, whose story we told, the rise of our view of the cosmos out of the augmented experiences of the human eye, by now carried far indeed.

The second chapter turns to change. To the eye change is all-encompassing. But tools as simple as sealed containers and the two-pan balance can retrieve within all change a constancy that was first seen by the artisans. The cycles of life as well are governed by such constancies. We applied one universal theory to gain insight into the premier cycling road race, the Tour de France. Such an experience is too complex for the lab. But even heroic human performance is demonstrably constrained and guided by the laws of change. New ideas and new precision unified change still more: Einstein's $E = M$.

In the third essay on mapping the world, another theme enters. The senses again find extension; but a problem that appeared to be a wholly visual and geometrical one—pointing at a distance—became entwined with the order of the stars and the fall of the plumb bob. Step-by-step approximations improved the map as understanding grew. Nowadays we can map not by direction at all but by time. Visual creatures that we are, we try to map experiences that are far from spatial into something we can see. The map of coasts and mountains becomes a metaphor for theory throughout science.

In the fourth example we entered field geology, where direct sensory experience still counts for much, and where cogent narrative is more central than mathematics. The hammer of the geologist turned into an ocean drillship; with that new instrumentation, but under the same style of inquiry, an astonishing past appeared. First suggested in a dazzling flash of recognition, then step by step, experience with gravels and dams led to a documented chronicle of a whole ocean suddenly turned desert and calamitously turned ocean again, in a swift sequence without parallel.

The last two chapters are tied together (a physicist-astronomer could not resist cobbling on his own last). First we examined how experiences wider than the work of goldsmith, cook, and lapidary all hinted at hidden atoms. Their size and number came clear as chemical experience grew and was reflected upon. The surprise of the quantum came from the study of the color of flame. It led to our present mastery, a single individual atom posed for its picture. Here it is the convergence of manifold experience that we emphasized, and the willingness to let experience lead even to "alterations in point of view," some of which are genuinely enduring.

That same mastery had assessed the recipe for the cosmos, read off in colored light. For fifty years the main result was steadily confirmed; the universe we see is mainly the simplest of atoms, hydrogen. But in the last decade doubt has entered with a fanfare of drums. The work was not wrong; all that we can see is hydrogen, all right. But we have found by remote weighing that what we see is only a tithe of what is in space. Most matter is entirely dark. What it is, how this can be, we have yet to learn. Doubt is as real a part of living science as is assurance.

Another part of the story is here as ambience. Science is a community, extended and diverse in time, place, origins, skills, personality. The places we visited in field and lab and museum and workshop, and the people who took part there, are visible too, warm in our memory. We could not visit all of science. Biology entered through perception, through the physiology of the athlete, through the fossil record, but hardly as it flourishes today. Someone else will have to go there.

Nor did we treat the sciences of the behavior of human beings and of their societies. I regard the social sciences as continuous with the natural sciences. The light passes through the telescope on to the eye, and understanding the perception requires understanding the astronomer, not only astronomy. That study is much harder, but it is growing, too. History plays a growing part in natural science, and we slowly come to understand the demands it makes on method and experience, more subtle than those of the repeatable flight of projectiles.

That, too, is a path for others to guide.

When gold coins were much used in commerce, experience suggested that the ring of the metal on the tabletop could exclude some base coins whose weight alone would let them pass. It was prudent and even satisfying to test for the ring of truth. Of course, that was not yet the truth; some counterfeits rang true. But it was an approach to the truth, which is about all anyone can expect.

We have examined a few of the many and growing ways by which we seek the ring of truth; without some way, without some coherent evidence of your own, you cannot judge well. Certainly in science it is less often deception than it is self-deception we must guard against; but in science and beyond, the best stance is to seek that clear ring. Once sought, once heard, we ourselves have changed.

Philip and Phylis Morrison
Cambridge, Massachusetts
3 March 1987

RESOURCES FOR THE READER

SOURCES AND NOTES

Title page   The figure of the lynx is taken from the title page of Galileo's 1613 book on sunspots, where it appeared as the emblem of the Academy of the Lynxes. By permission of the Houghton Library at Harvard. We colored the lynx in the old manner.

CHAPTER 1: LOOKING

opposite page 1   The green studio picture is a direct photograph; the other photos were made from video images in the video studios of Vizwiz, in Brookline, Massachusetts, where the entire scene was manufactured. The baseball pictures are with the permission of Major League Baseball Productions in New York City.

page 2   From René Descartes, *Tractatus de Homine,* 2nd edition, 1677, by permission of the Houghton Library at Harvard. The Houghton has generously allowed us access to their treasure of old and rare books. Many of the photographs from books that appear in our volume were caringly copied by Rick Stafford, Michael Nedzweski, and Elizabeth Gombosi, photographers at the Fogg Art Museum.

page 3   The drawing of the camera shutter and many more in this book were thoughtfully prepared by Al Jarnow, Northport, New York, who was also the animator for the films.

page 4   Photo of turning cake by Michel Chalufour, one of the film editors for the TV series.

page 5   Photographs of McBride magic made near Quincy Market, Boston, from the PBA film.

page 6   Mirror action at an exhibit in the Exploratorium, photo by Esther Kutnik.

page 7   Drawing of Kanizsa images from the Exploratorium.

page 8   Random triangle image from Bela Julesz in his *Foundations of Cyclopean Perception,* University of Chicago Press, 1971.

page 9   Cube photo by Parthiv Shah; drawing by Al Jarnow.

page 10   The Dürer scheme for drawing in perspective is from *Undermensung der Messung,* 1525.

page 11   The architectural drawings by Vreedeman de Vries are from his book *Perspective,* 1604. By permission of the Houghton Library, Harvard University.

page 12   This Ames Room was planned by William Walton at the Science Museum of Virginia, Richmond. Photo by Cyane Lowden.

page 14 and 15   The frescoes of the scholarly monks are in the Chapter House of St. Nicholas Church at Treviso. Photos by Foto Fini.

Friar Giordano was the priest from near Pisa who reported that he had met the inventor of spectacles. The sermon is cited on page 27 of a classical piece of scholarly detection, by the late Edward Rosen: "The Invention of Eyeglasses," in the *Journal of the History of Medicine and Allied Sciences,* Vol. 11, 1956.

pages 16 and 17   The bird's-eye view of Venice is a big wall map, about five feet by eight. It is reproduced by courtesy of the Trustees of the British Museum.

page 18   The glass furnace is from *De Re Metallica*, G. Agricola, 1556. From the collection of Cyril Stanley Smith.

Beads in the collection of the authors, photo by Michel Chalufour.

A few concise pages on the whole story of glass, carrying it on into China, are found in the appropriate volume of that treasury of the history of science and technology: Joseph Needham and others, *Science and Civilization in China,* vol. 4, part I: Physics, Cambridge, 1962.

page 20 and 21   The old lens-grinding device is from *L'Occhiale All' Occhio...*, Carlo Antonio Manzini, 1660. By permission of the Folger Shakespeare Library, Washington D. C.

The photos of lens grinding were made in the Dilworth basement shop by William Leatherman.

page 21   Photo of lenses in different stages; by Michel Chalufour.

page 23   Text seen through lenses: photo by Michel Chalufour.

page 24   Using Galilean lenses in San Marco Square, from the PBA film.

page 25   Church steeple taken in Concord, Massachusetts, by Michel Chalufour.

page 26   Murano as first seen through the telescope was described by Antonio Priuli, cited from his journal in *Le Opere di Galileo Galilei* vol. 19, p. 587, edited by E. Favaro. Translated from the Italian by Elizabeth Cavicchi. Scene from the PBA film.

Drawing of Galilean telescope by Al Jarnow. Not long afterward another form of telescope was invented by Johannes Kepler, and called by his name. It has two convex lenses of different curvature, the stronger of them near the eye. It usually has a larger field of view, but in earthbound use it has the fatal flaw of inverting the image. Astronomers do not mind that, and it is often called the astronomical telescope. Opera glasses and spyglasses are still Galilean.

page 27   We examined Galileo's telescopes with the indispensable assistance of historian Thomas Settle, Brooklyn College. Photo courtesy the Museo di Storia della Scienza, Florence.

Letter on using the telescope cited in *Galileo at Work,* by Stillman Drake, p. 147 (See More to Read)

page 28   Galileo's marvelous sketches of the moon are at the Biblioteca Nazionale Centrale, Florence. Reproduced with their permission.

page 29   These passages on the new world of the moon are from Galileo's first work on the telescope. The translation from the Latin is by Stillman Drake, in his collection *Discoveries and Opinions of Galileo.*

The portrait of Galileo is from the Biblioteca Marucelliana, Florence.

pages 30 and 31   The mosaic of the Creator regarding from the outside the luminous and circular work of His hands is found in the Cathedral of Monreale near Palermo; it was completed around 1200. Photo: Scala/Art Resource.

page 31   The astrolabe made and signed by Nastulus in the year 927-28 is the oldest among the thousand or two thousand we have of all its kind. It is part of the Al-Sabah Collection at the Kuwait National Museum.

page 32   One of the images of the moon that Galileo drew, a detail from the array of six moon images.

In 1579 the astronomer Michael Maestlin drew what he had seen when he looked at the Pleiades; it was reproduced in 1666 in *Historia Coelestis ex libris...Tichonis Brahe,* edited by L. Barettus.

Galileo's map of the Pleiades comes from his *Siderius Nuncius,* 1610. Reproductions by permission of the Houghton Library, Harvard University.

page 33   Sunspots from Galileo's *Istoria e Dimostrazioni...intorno alle Macchie Solari*, 1613. This reproduction by permission of the Houghton Library, Harvard University.

Drawing of Saturn from a letter of July 30, 1610. From the Biblioteca Nazionale Centrale, Florence.

Notebook sketches of the movements of the moons of Jupiter. Galileo's numbers measure the moons' distances in terms of Jupiter's disk radius (which he did not draw to scale). Consult Stillman Drake, *Telescopes, Tides and Tactics*, University of Chicago Press, 1983. The image is from the facsimile in *Le Opere di Galileo Galilei*, vol. 3, part 2, 1907, edited by E. Favaro.

Venus seen in three phases from a letter of Galileo dated February 25, 1611. From the Biblioteca Nazionale Centrale, Florence.

page 36   *The Immaculate Conception* by Murillo; Escorial, Prado. Photo: Scala/Art Resource.

page 37   Cigoli's painting in the Church of Santa Maria Maggiore, photo by Wolfgang Achtner.

page 38   The drawings of Saturn are by Galileo's rival, early telescopic astronomer Christopher Scheiner, in his *Disquisitiones Mathematicae*, 1614. By permission of the Houghton Library, Harvard University.

The planet drawings are from Huygens' 1650s manuscripts (ms. Hug. 5) held in the Bibliotheek der Rijksuniversiteit te Leiden.

page 39   The drawings of the craters on the moon and the Pleiades are from the *Micrographia* of Robert Hooke, done in 1665. Schema XXXVIII. By permission of the Houghton Library, Harvard University.

page 41   Earth satellite drawing and quotation from Newton's *System of the World*, in the English translation of 1728. By permission of the Houghton Library, Harvard University.

page 43   The amateur telescope makers at Stellafane photographed by Dennis di Cicco, who himself made the eleven-inch telescope in the foreground.

page 44   Photo of making the mirror for the space telescope, courtesy Perkin-Elmer Corporation, Hartford.

The Very Large Array on the Plains of San Agustin near Magdalena, New Mexico, a close view of several of its twenty-seven eighty-two-foot dishes: Courtesy National Radio Astronomy Observatory, Associated Universities, Inc.

page 45   The 200-inch in its dome, drawn by Russell W. Porter, 1938. Photo from Palomar Observatory.

page 46   The flashes of the Crab Nebula were discovered by John Cocke, Michael Disney, and D. J. Taylor of the University of Arizona using electronic pulse recording. The classic strobe views in our picture were made later in 1969 through the 2.1-meter telescope at Kitt Peak by Hong-Yee Chiu, Roger Lynds, and Steven P. Maran, using the "Pulsar Hunter" instrument developed by NASA-Goddard. Photos from National Optical Astronomy Observatories, Tucson, Arizona.

An image of the Crab Nebula taken through a polarizing sheet, made by Walter Baade with the 200-inch telescope in the late 1950s. Palomar Observatory Photograph.

page 48   Newton's spectrum from *Elemens de la philosophie de Neuton* by François Voltaire and his learned companion, Mme. du Châtelet, published in 1738. By permission of the Houghton Library, Harvard University.

The all-but-incredible tale of the origins of the magnetic compass among Chinese soothsayers 2,000 years ago is elaborated by Joseph Needham, in the same volume 4, part 1, cited earlier, pp. 229-79.

The photograph showing the response of thermometers to the infrared is by Wolf Rueckner at Harvard University. The same experiment shown in the old print was that of astronomer William Herschel, "Investigation of the Powers of the Prismatic Colours to Heat and Illuminate Objects," *Philosophical Transactions of the Royal Society*, London, 1800. MIT Archives.

page 49   We used a carbon arc and a big quartz prism to get plenty of ultraviolet in our spectrum. Demonstration by Wolf Rueckner, Department of Physics, Harvard University. Photo by Parthiv Shah.

X-ray photo of hardware in a box, made by G. S. Moler, Cornell University, January 1896; courtesy Burndy Library, Norwalk, Connecticut.

Radio map of the sky in the UHF band (256 Mhz) made with a ninety-six-helix radio telescope at Ohio State University by John Kraus. From his book *Radio Astronomy*, 2nd Edition, Cygnus-Quasar Books, Box 85, Powell, Ohio 43065.

page 50   Butterflies illuminated in sunlight, with lots of ultraviolet and all the visible light as well. Then the color image is normal, suited to our eyes and our color film. The bluish image is made with a filter that passes only the ultraviolet on to the lens, so that the camera records only the positions of ultraviolet color, even though all colors are present in the light. Photos by Tom Eisner, Department of Biology, Cornell University.

page 51   Spoiled tomato as a special favor from our scrupulous neighborhood grocery; photo by Michel Chalufour.

Bacterial colonies (strains of colon bacillus) growing on nutrient agar. Photos by Jack Goldstein.

page 52   The use of clear agar gel to solidify the growing surface was the idea of Frau Fanny Hesse from New Jersey, who was one of the circle around the great pioneer bacteriologist Robert Koch in Berlin during the 1880s. The year 1987 is the centennial of the petri dish itself: time to commemorate Richard J. Petri. See Ronald Atlas, *Basic and Practical Microbiology*, 1986.

pages 53, 54 and 55   Leap-frogging gentlemen from Muybridge's *Animal Locomotion*, Plate 169, 1887. Van Pelt Library, University of Pennsylvania, Philadelphia.

page 54   Strobed photograph of the dancer Nora Kaye, by Gjon Mili. Life magazine, © Time Inc.

page 55   Nude Descending a Staircase #2, Marcel Duchamp. Philadelphia Museum of Art: The Louise and Walter Arensberg Collection.

CHAPTER 2: CHANGE:

page 56   A. Einstein, cyclist, Pasadena, 1931. Courtesy of the Archives, California Institute of Technology.

pages 58 and 59   Cyclists of the Tour de France. Photo by Presse-Sports.

page 58   Popcorn strobe: flashes at 20-millisecond intervals by Bruce Dale and Gregory A. Dale.

Forest fire: in Ocala National Forest, Florida. U.S.D.A., Forest Service.

page 59   Volcano: Johnston Cascades Volcano Observatory, Vancouver, Washington. U.S. Geological Survey, photo by Lyn Topinka.

Sunflowers: growing in Minnesota. Photo by Donald S. Dean, Berea, Ohio.

Rust pile at a junkyard: photo by PM.

page 60   A sparkler and its dust: photos by Michel Chalufour.

page 62   Matches on a balance: photos by Michel Chalufour.

page 63   A test tube with mercury and its red oxide: photographed by Chip Clark, Washington, D.C.

page 64   Vessel to catch gases from Priestley's treatise *Experiments on Air*, vol. II, 1775. MIT Archives.

pages 65 and 66   Tom Tompkins and his assistant Thomas Fulghum built the lively sealed box. Photos at the Exploratorium by Parthiv Shah and Ron Scherl. The Scherl photos are copyright by Ron Scherl.

page 68   Balance room of Carol Westmoreland at ERT/A Resource Engineering Co., Bedford, Massachusetts. The weighing element of this most delicate balance is a fine quartz fiber that would bend under load, except that it is held almost in place by an automatically adjusted and measurable electrostatic counterforce. The balance is distributed in the U.S. by the Brinkmann Instrument Company and is made by Sartorius GMBH in Göttingen, GFR. Photo by PM.

page 70   The sphere of the earth is a NASA photograph.

The sealed glass sphere is made by EcoSpheres Associates in Tucson. The plants growing in it are green algae. The animals are tiny red Pacific shrimp; they live for five years or more, but they do not breed in so small a space. The bacterial soup has many unseen species. The oldest little sealed ecosphere (no shrimp) is still a living world after some fifteen years. The photo is by Parthiv Shah.

page 72   A racer of the 1986 Tour de France, photographed by Marian White.

page 73   Map of the Tour by Al Jarnow.

page 76   Chef Jean Colletaz and his staff of the Hotel Novotel outside Toulouse prepared the food we display here and welcomed us to his kitchen. Photos by PM.

page 77   The menu list is from Dr. Jack Harvey.

page 78   We imported into France, a little anxiously, an ample supply of all-American jelly doughnuts from Mr. Donut in Woburn, Massachusetts, and burned them as a bonfire one summer night on the beautiful farmland of our host M. Alberto Pellegrino at Lias. They were freeze-dried by Bob White of MLG Labs, Hingham, Massachusetts. Photo by PM.

page 80   The formal metric unit of energy is the joule. One kilocalorie is 4.19 kilojoules.

page 82   Wind tunnel photo by Dölf Preisig, Ringier Dokumentationszentrum, Zurich.

Photo of the Tour de France by Alexander Hubrick. Copyright Alexander Hubrick/The Image Bank.

page 83   The graph of bicycle speed records is from C. R. Kyle's piece (see "More to Read") in *Scientific American*, December 1983. Redrawn by Al Jarnow.

Motion can be transformed to heat, and hence directly to the calories we need. Long ago the simplest of mechanical changes was carefully converted to a measured amount of heat. A heavy weight on a cord slowly fell, turning as it went a big paddlewheel within a heat-insulated tank of water. Temperature rise gave just how many calories of heat were made in the mechanical process of allowing a standard weight to fall a measured distance under gravity. The effect of motion under gravity was then reckonable in calories: the mechanical equivalent of heat.

page 85   Steve Ball's vehicle Dragonfly, photographed in 1983 by David Gordon Wilson, MIT.

Three dozen short test flights of the prototype Daedalus aircraft were made during 1986 at Hanscom Field, Bedford, Massachusetts. The pilot here is Lois McCallin. Photo by Peggie Scott.

page 88   Boiling kettle: photo by Michel Chalufour.

page 89   The cycle-airplane takes off early on a December morning. Photo by Peggie Scott.

pages 90 and 91   Strobe photos of a two-inch windup jumping kangaroo, about tenth-second intervals, by Papa Flash himself, Dr. Harold E. Edgerton, Jr., of MIT.

page 93   Our guides to the glowing wonder were C. J. Goebel and Wayne Amos, at Mounds. Photo made in the 1970s at Mounds Laboratory, Monsanto Research Corporation, Moundsville, Ohio.

CHAPTER 3: MAPPING

page 96   The Lane pocket globe and case: photo by Michel Chalufour.

page 98   The Eastern U.S. and Europe by night. A pair of satellites of the U.S.A.F. Defense Meteorological Satellite Project routinely take their strip scans day and night from 500 miles up. Their mosaic images show moonlight and the aurora, as well as most other fires and lights of nature and man. Produced from USAF DMSP film transparencies archived for NOAA/NESDIS at the University of Colorado, CIRES/National Snow and Ice Data Center, Campus Box 449, Boulder, Colorado 80309.

The whole earth. Meteosat images track the weather for Europe, here on April 8, 1979. Courtesy European Space Agency, Paris. Copyright ESA/METEOSAT.

page 99   Japan. The lights of an isolated ship do not appear, so the oceans are usually dark. But one night in the Sea of Japan more than a thousand boats gathered close, most of them burning one or two hundred kilowatts of incandescent lamp power. "...it must seem like daylight out there." See Thomas A. Croft, *Scientific American,* July 1978, p. 94.

page 100   Map from the very popular German atlas of the turn of the century, *Handatlas,* by Adolf Stieler, about 1892. From the MIT Library. Photo by Parthiv Shah.

page 101   Jane Bradick painting, *View of West Front of Monticello and Garden,* photo courtesy of Thomas Jefferson Memorial Foundation.

Map by Thomas Jefferson Randolph, in 1808. Plan from the Massachusetts Historical Society, Boston.

Photo of Monticello and our mapping party, from the helicopter by Peter Hawkins.

page 102   Amiable adviser and chief of our party (and owner of the plane table) was the experienced planetabler Fred Wiseman of MIT. Photos of our day's surveying work by Jack Goldstein, professor of physics, Brandeis University. As insightful photographer and careful reader of the images and text, he has helped this book in many ways.

The drawing from George Adams's *Geometrical and Graphical Essays,* 1791, shows one of his simpler instruments, though somewhat more complicated than ours. By permission of the Houghton Library, Harvard University.

page 103   You could make such a map yourself, indoors or out.

pages 104 and 105   Passing landscape sketches by Al Jarnow.

page 105   This diagram of mapping accompanies a problem for students using George Adams's *Geometrical and Graphical Essays.* By permission of the Houghton Library, Harvard University.

page 106   The references present a good summary story of the great survey of India. The image here is from *The Great Trigonometrical Survey of India,* vol. 1, J. T. Walker, 1870. The altitude of Everest has been very well confirmed by now, and is known to within the variations in its snow load, say some tens of feet.

page 107   Land-use details from aerial photos of the Wellsville, Ohio, Quadrangle. Aerial Mapping Photography, U.S. Geological Survey.

page 108   Orion moves across the sky. Photos made in Texas by Dennis di Cicco. He took a long series one night using a Minolta X700 camera with programmed back to control timing and exposure on Fujichrome 400D. The lens was a 16mm fisheye, f2.8.

page 109   Drawing of Orion from a manuscript found in Tunhuang by Sir Aurel Stein. Photo from the British Museum.

page 110   Highway map by Rand McNally. Copyright License number RL-87-S-62.

Sightings to a star with a portable instrument: geometry by Euclid, diagram by Al Jarnow.

page 111   The sturdy Van of Eratosthenes was lent us by Kurt Skallerup, of Ryder Truck Rentals, Omaha, Nebraska.

page 112   Dennis di Cicco made a series of sky photographs from our post near Bassett the night we began to measure the earth. He used the camera described in the endnote on Orion, but with an ordinary 28mm f2 lens. Arc drawn by Al Jarnow.

page 113   Road straight across the hills, scene from the PBA film.

page 114   Van photo by PM: drawing by Al Jarnow.

page 115   Earth sphere showing the equator and a parallel of latitude, drawn by Al Jarnow.

page 116   Ptolemy's second-century data mapped from *Opus Geographiae*, L. Phrisius, editor, Strasburg, 1522. The Houghton Library, Harvard University.

page 117   The earth is not perfectly spherical, but flattened by a percent or so at both poles. (Ocean floors and mountains are a part in a thousand.) A trip would show that up as slightly different road distances for a given shift in star direction as one moved north, but both the effect and the calculations are delicate. A nonspherical but spinning earth has a complicated effect on the direction of the vertical plumb bob, which no longer points straight to the center of the spheroid. Now both geometry and gravitational physics are needed. The method permits its own check; with enough careful measurements you can plot the whole shape of earth, if you are willing to travel.

page 118   Not in Kansas; photo by PM.

page 119   The instruments we brought to Ellicott's Hill: photo by PM.

Jim Moss's clock was made to a design of the late eighteenth century, a good timekeeper, but using an escape-ment less advanced than Ellicott's. Scene from the PBA film.

American-made Keuffel & Esser transit of some fifty years age, lent by Fred Wiseman. Note circles and leveling screws. The expert user of the transit was Thomas Hendley, Natchez surveyor with Jordan, Kaiser and Sessions, who has spent much time retracing the old surveys of Andrew Ellicott and his successors in this part of the country. Photo, PM

Reprise of the Lane globe photograph from the opening page of this chapter.

page 120   The telescope and the equal-angle instrument are ones that Elliott brought to the hill in Natchez. The highly accurate large pendulum clock, very similar to the one Ellicott used, is one made by his friend David Rittenhouse before 1769. Ellicott's instruments are held at the Museum of American History, Smithsonian Institution, Washington.

The Rittenhouse clock photograph by courtesy of the American Philosophical Society, Philadelphia.

page 121   Star dial drawn by Al Jarnow. The twenty-four hours of a sidereal day are four minutes shorter than the familiar mean solar day, so star-time days, hours, minutes and seconds are all proportionately a little shorter than those in everyday use.

page 123   Sequence of finding Natchez drawn by Al Jarnow.

The tables we used are in *The Astronomical Almanac* for the year 1986, an annual jointly published by the U.S. Naval Observatory and the Royal Observatory of Greenwich. Mr. Ellicott used tables both from Paris and from Greenwich.

page 124   Map of old Natchez from the book of a French traveler, Victor Collot, *A Journey in North America…*, Paris, 1826. By permission of the Houghton Library, Harvard University.

page 125   Photos of Jupiter and his moons, by Dennis di Cicco, taken at his eleven-inch Celestron telescope in 1986, at Sudbury, Massachusetts. He used special optics to obtain a very long focal length for the images; he shot at five-minute intervals, but we selected only a few of the images. You would see other work of his, along with that of many others, in *Sky and Telescope* magazine.

page 126   The famous engraving of the King's visit to his scientists is by artist Sebastien Le Clerc, himself a scientist of the time. The image is described in the annotated frontispiece of Singer's *Short History...* (see "More to Read"). It appeared in this form in 1676 in Denis Dodart's *Mémoires pour Servir a l'Histoire des Plantes.* By permission of the Houghton Library, Harvard University.

page 127   Below a photograph of Jupiter with his moons are the tables Cassini published, *Ephémérides,* in 1668, Bologna. We have laid two of his pages end to end and have drawn on the photo a curve for the motion of one satellite to suggest how predictions might be made. (Galileo had even made a brass model of the orbit circles for his early calculations.) By permission of the Houghton Library, Harvard University.

page 128   The comparison of the maps of France, old and new, was published by Philippe de la Hire and Jean Picard in 1693 in Paris. It appeared in *Recueil d'Observations.* By permission of the Houghton Library, Harvard University.

page 131   The published map of the world, the Planisphere Terrestre, has been reproduced by the Clements Library of the University of Michigan. The places of reliable position that are marked on it by an asterisk are listed in the fine little book of L. Brown cited in "More to Read." We have marked them in red.

page 132   San Francisco control point photos by Parthiv Shah.

page 133   Benchmark No. 10 of the San Francisco Harbor series is entered on this topographic map of the city, along with another dozen or two control points of several degrees of reliability. Its measured position is thought to be accurate to fifty or a hundred feet on the map of the world, about a part in a million of the global dimensions. U.S. Geological Survey map.

The widely available maps of the United States Geological Survey are a national resource in all their variety and utility. We could hardly imagine work without them, and commend them to stay-at-home readers and travelers alike. The people of the U.S.G.S. have been of great help to us, at all four centers, in particular the topographic mapping group at Menlo Park, California.

Our drummer on the MIT soccer field, using a Harvard University Band drum, was the Boston percussionist-composer Joyce Kauffman.

page 134   Depth meter drawn by Al Jarnow. Our ship was Amphitrite, Captain H. Arnold Carr. The depth meter we used was loaned to us by Lowrance Electronics, Inc.

page 136   Using a portable terminal for the GPS system. It is made by ISTAC, Inc., Pasadena, California, and is here operated by its developer, Peter MacDoran, and engineer Al Buennagel. Photo by PM.

page 137   How GPS works: drawing by Al Jarnow.

page 138   The computer printout of the ISTAC GPS program was made in the backyard of the old tavern on Ellicott's Hill, Natchez, in 1986.

page 139   GPS results for relative distances, say for the two ends of a survey line, can be good to a part in a million, inches out of a hundred miles. Drawing by Al Jarnow. A recent tentative satellite-assisted survey threatens Mount Everest's record, not by shrinking its height, but by increasing the height of the runner-up, peak K-2, by nearly a thousand feet. Other new GPS results are stirring around all the USGS control points in the United States, of course only by a hundred feet or so.

page 141   Computer summary of data on Io eclipses. The first shows twentieth-century eclipses, while the second adds to that a summary of all the reports of eclipses of the first moon, Io, back to the mid 1600s. We drew the red trend line. From Jay H. Lieske, Jet Propulsion Laboratory.

pages 142 and 143   These are images in enhanced color of the four Jovian moons first found by Galileo. Here they are as they appeared when they were first made into physical—and quite distinct—places in early 1979 by the flyby of the video cameras on Voyagers 1 and 2. Their relative sizes are correctly shown. Our own moon fits by size between Io and Europa. Jet Propulsion Labaratory photos.

page 144   A part of the videotape that mapped the Music Mountains, the work of computer scientist Peter Roos, New York City.

page 145   This solid is a version of the color theory of A. H. Munsell, based on many matching and blending experiments using colored surfaces. He named the three color dimensions hue, value, and chroma. The Munsell system goes back to the early years of the century, and is still in wide practical use. Colored light measurements rather than pigment colors are also much used for color-matching and color reproduction. The figure comes from Howard T. Fisher, the Laboratory for Computer Graphics and Spatial Analysis, Harvard University Graduate School of Design, 1971.

CHAPTER 4: CLUES

page 146   The Yellowstone geysers are watched closely by a set of amateurs who have, for example, named and chronicled the state of thirty-odd geysers in this northern portion of Norris Basin. See T. Scott Bryan, *The Geysers of Yellowstone,* Colorado Associated University Press, 1986. Photo by PM.

page 147   Geyser plumbing is nearly always contained in rocks with a quartz component. The hot waters dissolve the silica and may redeposit it at cooler surfaces. That is the basis for the inconstancy and competition among geysers.

page 148   The color photos were made by Richard Yuretich on his trips up the slopes.

page 149   The two old photos are by J. P. Iddings, U.S. Geological Survey: the single tree trunk was published in U.S.G.S. Monograph 32, part 2, 1899; the other, on Specimen Ridge itself, in U.S.G.S. Folio 30, Fig. 2, 1896.

Pencil sketch of Amethyst Mountain by W. H. Holmes, done about 1879, from the National Archives.

page 151   The great ash cloud from Mount St. Helens: photo by Gary Rosenquist, Earth Images, Bainbridge Island, Washington.

Trees after the Mount St. Helens eruption: photo by Lyn Topinka. U.S. Geological Survey, David A. Johnston Cascades Volcano Observatory, Vancouver, Washington.

page 152   A cut cake: photo, Michel Chalufour.

page 153   Old drawing of the forest layers on Amethyst Mountain by W. H. Holmes, published in the *Bulletin* of the U.S. Geological and Geographical Survey of the Territories, vol. 5, no. 1, 1879.

page 154   Glomar Challenger Leg 13 map and route, drawn by Al Jarnow.

page 155   Photos of the research drillship *Glomar Challenger,* Deep Sea Drilling Project, Scripps Institution of Oceanography, U.C.S.D., La Jolla, California.

pages 156 and 157   The sea-bottom images come from the British Challenger Office's *Report on the Scientific Results of the Voyage of HMS Challenger* (18 73-6) The ooze is plate XI of the Deepsea Deposits volume. The radiolaria are plate 81 of volume 18 (Radiolaria 1887). From the Harvard College Library.

page 158   Compensation for the motions of sea and wind while the ship is drilling is a major requirement. Two miles of drill stem may bend and give a little, but the ship must stay reasonably aligned while the bit below is turning in the hole. This is accomplished dynamically. The ship carries four side thrusters, powered small propellers able to push it to and fro in quick response. They are instructed by a computer, which monitors at the corners of the ship the delay of acoustic signals that come up from a special beacon dropped to the seafloor. A drift of about 200 feet is tolerated, the drill stem bent something like a degree.

page 159   Three drawings of the business end of the drill stem, by Al Jarnow.

page 160   Shipboard snapshot, Deep Sea Drilling Project, Scripps Institution of Oceanography.

Maria Cita: photo by Ken Hsü.

page 161   Seismic echo profile of M-reflector from Bill Ryan, made on research ship Robert D. Conrad, 1965. These profiles presented the puzzle the geologists were trying to understand. Lamont-Doherty Geological Observatory, Columbia University.

pages 162 and 163   The gravels from below the Mediterranean's bottom: photos by Michel Chalufour.

page 165   Photo of the great Hoover Dam by E. E. Hertzog, Bureau of Reclamation, Boulder City, Nevada.

page 166   Mediterranean photo from Meteosat, the European Space Agency's geosynchronous satellite. A volcanic plume from Mount Etna on Sicily can be made out, well east of Gibraltar. Meteosat image from the European Space Agency. Copyright ESA/METEOSAT.

page 167   LANDSAT image of Lake Mead, December 1982.

page 168   Drawings of the drying Mediterranean and its lakes by Al Jarnow.

page 169   Mosaic photographs of a drying puddle by Al Jarnow.

page 170   Bill Ryan in the core library at Lamont-Doherty Geological Observatory. Photo by William Leatherman.

page 171   These are cores 13, 9, and 8 from site hole 124 of leg 13 of the Deep Sea Drilling Project, its first Mediterranean cruise (see map). Lamont-Doherty Geological Observatory, Columbia University.

page 172   The ocean in a bowl dries up: photos by Ed Joyce, The Frame Shop, Newton, Massachusetts.

page 175   The Nile sectioned at Aswan, redrawn by Al Jarnow. The subsurface profile is from I. S. Chumakov, Geologicheskiy Inst., Akademiya Nauk SSR, Moscow. English abstract and figure in *Reports of Deep Sea Drilling Project,* vol. 13, part 2, 1973: Scripps Institution, U.C.S.D.

page 176   Maria Cita speaks for the film in her lab in Milano.

page 177   The boundary between the two geologic epochs signals the end of the brief interval of the desert. Above it the deep ocean had rolled in. It has never left this spot. The marine fauna of today's Mediterranean are the same as in the Atlantic, as they have been for about five million years. That abrupt boundary now has a context, the refilling of the sea. But the boundary itself has been recognized since the early nineteenth century, when Sir Charles Lyell made it the defining interface between two of the divisions of the worldwide era of recent life, the Pliocene above and the Miocene below.

Frederick Edwin Church: *Niagara,* painting in the collection of the Corcoran Gallery of Art, Museum purchase.

page 178   The temperature effect is a subtle one upon the chemical reactions of growth; the ambient temperature determines the relative rate of incorporation of two isotopes of oxygen within the material of the shell.

Dialogue on the ridge of the Rock of Gibraltar, 1986.

page 179   Al Jarnow imagines the ancient Falls of Gibraltar.

CHAPTER 5: ATOMS

page 180   On the Charles River, July 4, 1987: photos by Michel Chalufour.

page 182   Fireworks photo by Parthiv Shah

page 183   Spinthariscope photo by Michel Chalufour.

To see flashes in the spinthariscope, the dark-adapted eye is a fine low-level light detector; it was not so easy to present those flashes second-hand.

For photos, the atomic (alpha-particle) flashes on the spinthariscope screen were sampled for a tenth of a second at times tens of seconds apart by a CCD (charge-coupled device) imaging system, and the photos were produced electronically from the CCD record. Edward Dunham of MIT was the master of this up-to-date sensitive system.

It was even harder to show the faint flashes in motion on television. The image intensifier and sensitive video camera used were generously tried out for our job and then loaned to us by the manufacturer, Electrophysics of Nutley, New Jersey.

Mr. Billmeier's words were spoken to camera for the film.

page 185   Gold-beating scenes from the PBA film.

page 186   Translucent gold: photo by Michel Chalufour.

page 187   The assay lab ornaments the title page of the *Treatise on Ores and Assaying*, Lazarus Ercker, Prague, 1574. From the collection of Cyril Stanley Smith, who made a translation of this book with Anneliese Grünhalt Sisco, University of Chicago Press, 1951.

page 188   Cupels being made, from the *Treatise on Ores and Assaying*, Lazarus Ercker, Prague, 1574.

Assaying and glowing cupel: scenes from the PBA film.

Stained cupel and gold bead: photo by Parthiv Shah.

page 189   Photos of a wide beach and of winnowed sands by Parthiv Shah at Crane's Beach, Ipswich, Massachusetts.

Photos of that sand under a lens, Michel Chalufour.

page 190   Magnet and sand photo by Michel Chalufour.

page 191   The burnt dinner: photo by Parthiv Shah.

Silicon in such single-crystal wafers is routinely purified of other atoms to the level of parts in ten billion, better than 99.9999999 percent pure. Photo provided courtesy Digital Equipment Corporation, 1987.

The figures of the chapter present quite a few chemical elements in reasonable states of purity: carbon, gold, hydrogen, mercury, oxygen, silicon, silver. A few more elements are observed by specific light from glowing gas, with other elements present but unseen: barium, copper, rubidium, sodium, strontium, neon, argon.

page 192   Water electrolysis experiment by Tracy Jones, Department of Chemistry, MIT; photo by Parthiv Shah.

Water synthesis experiment by Wolf Rueckner, Harvard. Photo by Parthiv Shah.

page 193   Geodes from a Mexican streambed. Photo by Michel Chalufour.

page 194   Alum crystals grow from water solution. Crystals grown by PM, microcinematography by Boyd Estus in the PBA film.

page 195   Various crystal forms. The arrayed globules are from Scheme VII of *Micrographia,* by Robert Hooke, London, 1665. By permission of the Houghton Library, Harvard University. (The tiny crystals of his Fig. 1 he found within a cavity in flint. The objects rather like slate in his Fig. 2 are particles found in urine, seen by Hooke as magnified about 100 times.)

The modular crystal drawings, the first thing of their kind, are from *Trait de Minéralogie,* vol. 5, by René-Just Haüy. Paris, 1801. MIT Archives.

page 196   The process used to crystallize table salt varies; try a few brands to see which is producing cubes at any time. Photo by Michel Chalufour.

Pyrite: Mineralogical Museum, Harvard. Photo by Parthiv Shah.

A natural diamond crystal: Mineralogical Museum, Harvard. Photo by Parthiv Shah.

page 197   Cut and polished gem diamond in stand, from the PBA film.

We burnt the diamond in the vacuum furnace in the lab of Robert L. Coble, Materials Science, MIT: photos by William Leatherman.

Diamond char: photo by Parthiv Shah.

pages 198 and 199   Kin Jing Mark, master chef, is a vocational teacher at the China Institute, New York. Scenes of noodle-making are from the film.

page 200   Oil slick photos by Parthiv Shah.

page 201   Brownian motion: the position in the microscope of a half-micron resin particle in water, recorded every half minute. From F. Perrin, *Les Atomes,* Paris, 1913 (often reprinted). Redrawn by Al Jarnow.

page 202   These striking images of atomic arrays were made through a high-resolution electron microscope (model JEM-4000 EX), with magnifications of about three-quarters of a million in the microscope, the rest in the video system that displays the moving images. Videotapes from Dr. David J. Smith, Center for Solid State Science, Arizona State University, Tempe, Arizona.

page 203   Wire frame shadow photo by Michel Chalufour.

page 204   Raised gold cup, height 6 inches. From the royal graves at Ur, about 2600 B.C. Photo courtesy of the University Museum, Philadelphia.

The Jaguar figure of hammered and soldered gold is from the Moche culture of Peru, 400 B.C.-100 A.D., 5 inches long. Alsdorf Foundation and Primitive Art Purchase Fund, 1970.420: photograph © 1987 The Art Institute of Chicago. All Rights Reserved.

Fibula (safety-pin), 3 inches. Etruscan granulated work, about 650 B.C. Museum of Art, Rhode Island School of Design, Museum appropriation.

The granulated gold bead is the work of Noma Copley and Jean Stark, New York. Photo by Michel Chalufour.

page 205   The golden figure is from the Calima gold-working area, 300 B.C. to 1000 A.D.: Museo del Oro, Bogotá, Colombia. Photo by Jorge Mario Múnera.

Perpetual calendar with gold-leaf eclipses, a "girdle book," about 1461. Crawford Library, Royal Observatory, Edinburgh, Scotland.

pages 206 and 207   Flame photos by Wolf Rueckner, Department of Physics, Harvard.

page 206   Glowing wire in flame: photo by Parthiv Shah.

Rubidium flame and matching discharge tube: photo by Parthiv Shah.

page 208   Flare photo with grating by Boyd Estus. An optical diffraction grating is a set of grooves, equally spaced and parallel, precisely ruled on a transparent or reflecting surface. The spacing between the lines is comparable with the distance between crests of the light wave; the inexpensive grating used here has 13,500 lines per inch on a transparent plastic sheet, made by embossing the softer plastic by a ruled metal surface.

page 209   Lightning and its spectrum; photo by William Bickel, Department of Physics, University of Arizona, Tucson.

Spectra photos by Wolf Rueckner, Harvard.

page 210   Lights of a car lot at night: photo by Boyd Estus.

Spectrum with fluorescent screen: photo by PM.

page 211   Flower photos by Thomas Eisner, Department of Biology, Cornell University.

page 213   Toy atom energy states and spectrum, drawn by Al Jarnow.

page 214   The mercury light source for our spectral dissection, usually called the Franck-Hertz experiment after the first physicists to perform it in the years just before and just after World War I: photo by Parthiv Shah.

For our Franck-Hertz demonstration: the lamp in its housing, the variable interference filter device, and the photomultiplier detector: photo by Parthiv Shah.

For a variety of reasons, our experiment exploited the fast time response of the electronic detector. The energy of atomic collisions was controlled by an adjustable high-frequency current in a coil around the lamp that stirred up electrons inside to collide with the mercury atoms. The electrons came boiling off the filament within the lamp. Control was best while the filament was not adding electrons, during one half of the cycle of the alternating current that heats the filament. The photodetector was electronically blanked during the time the electrons entered.

The apparatus was developed and built by Scott Saleska, MIT physics senior, under the supervision of my friend and MIT colleague John King. We all acquired a still higher appreciation of the qualities of the quantum pioneers of fifty or sixty years ago!

page 215   The record of our spectrum-dissecting experiment, with additional drawing by Al Jarnow.

page 216   The energy states of mercury, drawn by Al Jarnow.

page 217   The chemist's letter appears in *Niels Bohr: A Centenary Volume*, Chapter 5.

page 218   To control our single atom, three levels of the barium quantum ladder are all involved. Both laser colors are on at once. The ion jumps easily to the uppermost state by absorbing light from the blue-green beam.

But it rarely drops down to the middle state from the top, usually going all the way down. Once it does enter the middle state, the red beam soon drives it back up. Thus the ion is caused to jump back and forth between lowest state and top state over and over again, each jump picking up or giving out the blue-green light. A recent account is given in *Science*, October 3, 1986, p. 24, by Arthur L. Robinson.

page 219   Hans Dehmelt's lab, University of Washington. Scene from the PBA film.

page 220   The unique photo of a single atom in color, the first one ever made, is the work of Warren Nagourney of the University of Washington research team.

He describes what he did: "Photograph of a single, cold barium ion confined in a radiofrequency trap to a region of space whose linear dimensions are less than 1 micrometer. The ion is illuminated by two lasers (red and green) which provide the cooling and allow the ion to be observed by eye, film and photomultiplier. This photo was taken with a Nikon F2 using a Micro-Nikkor 55 mm f/2.8 lens stopped down to f/5.6 with a magnification of about 5. The film was Kodacolor VR-400 and a 2 minute exposure was used. By replacing the camera body with a 20× microscope eyepiece, the ion is readily visible to the eye; it appears as a star-like point of bluish light."

CHAPTER 6: ASSURANCE AND DOUBT

page 222   Objective prism color plate of the Pleiades, Warner & Swasey Observatory, Case Western Reserve University. Photographed by astronomer Richard E. Hill, 1986.

page 224   This first printed star atlas of the whole sky is *Uranometria*, by Johann Bayer, Augsburg, 1603. The engraver was Alexander Mair of Ulm. Photo by Michel Chalufour.

The Pleiades on a print from the Palomar Observatory Sky Survey. In that great survey, most of the sky is shown twice on a couple of thousand big prints from the 1950s, once red-sensitive, once in the blue. Stars and galaxies are to be seen easily by the tens of millions, mostly as very small spots. This is a red print; the blue print shows this cluster as very nebulous and dusty. The thin streak along the lower edge of the photo is no scratch. It is the record of a meteor that passed by chance during the long exposure. Copyright © 1960, National Geographic Society-Palomar Sky Survey. Reproduced by permission of the California Institute of Technology.

page 225   A view of the Pleiades, photo by Dennis di Cicco.

page 226   Enlargement of a star's spectrum, seen first in color and then presented in black and white. If this had been made with the more sensitive black-and-white film usually used by astronomers, more spectral lines could be expected. Richard E. Hill, Warner & Swasey Observatory, Case Western Reserve University.

Richard Hill with the prism for the telescope, scene from the PBA film.

Objective prism drawn by Al Jarnow.

page 227   Nancy Houk at work in a scene from the PBA film.

Comparing spectra, in a scene from the PBA film.

page 228   A portion of one of A. J. Cannon's plates, at about its true size, from the plate stacks at Harvard College Observatory: photo by Harvard College Observatory.

page 229   Annie Cannon at work: photo from Harvard College Observatory.

page 230   Staff in 1925: photo from Harvard College Observatory.

page 231   Buildings at the observatory: photo from Harvard College Observatory.

page 232   The marked spectral plate used by Cecilia Payne is C15698 from the plate library at the Harvard College Observatory; the star whose spectrum you see is 18 Comae Berenices. Photo by Parthiv Shah.

Family snapshot of Cecilia Payne, from her daughter, Katharine Haramundanis.

page 233   Spectra drawn by Al Jarnow after photographic spectra from Nancy Houk and Michael Newberry's earlier spectral atlas, *A Second Atlas of Objective-Prism Spectra*, Ann Arbor, 1984

page 235   Calcium and hydrogen energy levels, drawn by Al Jarnow.

page 236   Abundances of atoms, in a compilation by the authors. Drawn by Al Jarnow.

page 238   The hydrogen gas of the Tarantula Nebula is lit by the fierce light of a brilliant cluster of hydrogen-rich stars within the Large Cloud of Magellan. Photo from the National Optical Astronomy Observatories.

The surface of the sun filtered to record only the light of the red hydrogen line. Photo National Optical Astronomy Observatories.

page 239   Saturn, mostly hydrogen, from a Voyager photo. Jet Propulsion Laboratory photo.

The galaxy NGC 4051 in a photograph carefully made in natural color by astronomer James Wray.

page 240   Photo of microwave horn about 1951: courtesy of Harold Ewen.

page 241   Photo of his electronics: courtesy of Harold Ewen.

page 242   George Carruthers in 1986 in a scene from the PBA film.

page 243   Ice cream weighing: photo by Michel Chalufour.

page 246   A supernova appeared in the spring of 1979 in the lower arm of the galaxy M100. The comparison picture was taken after the new star had faded. Such a supernova is close to a billion times brighter than a typical star like the sun; it stands out as an individual star even at a distance of tens of millions of light-years, and reveals the starry nature of its spiral patch of blue light, a galaxy.

The color photos were made on chilled film by an advanced amateur, Ben Mayer of Los Angeles, using a fourteen-inch telescope at his Problicom Observatory in the mountains northwest of the city.

page 247   All four of these variously tilted circular galaxies are seen in Lick Observatory photographs. The Andromeda galaxy, our closest full-size neighbor among the galaxies, is about 2.5 million light-years away. In clear dark skies it is visible to the unaided eye.

page 248   About galaxy names: the Andromeda galaxy = M31 = NGC 224; M101 = NGC 4457; M99 = NGC 4254; NGC 1087 has only that name; there are many other lists of various size and purpose, but most galaxies live unnamed by us, identified only by their positions.

page 249   These four images, all true to the apparent size of the galaxies as we see them, come from plates of the magnificent most-of-the-sky atlas of the Palomar Observatory Sky Survey, the source of our modern star-crowded photo of the Pleiades. The closest of the four is M101, about ten times as far as the Andromeda galaxy. Copyright © 1960, National Geographic Society–Palomar Sky Survey. Reproduced by permission of the California Institute of Technology.

page 250   Drawing of the cannonball shot round the world; from Newton's manuscript, by permission of the Syndics of Cambridge University Library, Cambridge, England.

page 251   Newtonian graphs for weighing the sun by using the motions of the planets: drawings by Al Jarnow.

page 253   The Doppler tonemaker was an audio oscillator built in a coffee can by the Helitron Company of Boston. The athlete was James Russell; his photo is by Parthiv Shah. The drawing of the motion and the shift in pitch is by Al Jarnow.

page 254   At the consoles of the Mayall Telescope on Kitt Peak in 1986, scene from the PBA film.

page 255   Kitt Peak vista: photo from National Optical Astronomy Observatories.

page 256   Video image of galaxy NGC 7541 made with a CCD camera at the Kitt Peak thirty-six-inch telescope by Judy Young.

page 257   Galaxy NGC 7541 shown both as a blotchy bright ellipse in the sky and as a spectrum. The spectrum is a photo made from an image intensifier at the spectrograph of the Kitt Peak four-meter telescope by Vera Rubin and W. Kent Ford.

page 260   Three graphs weighing the galaxy NGC 7541: prepared by Vera Rubin.

We thank our executive producer, Michael Ambrosino of PBA, Hugh O'Neill, editor, and Robert Aulicino, Art Director, Random House; no book without them!

THE BOOK OF THE FILM

Once upon a time a grade-schooler was taken to the movies in style, something beyond the customary kids' Saturday afternoon in the old Manor Theatre. A first-run silent film was showcased downtown, and Ramon Novarro or Rudolph Valentino would appear for two hours, without speech but with plenty of organ music, amidst Roman splendor or infantry action. As you left you always walked past men who announced urgently that they would sell the smooth paperbound brochure they called "the book of the film."

This volume, too, is "the book of the film." The film is, of course, the six-hour TV series shown on PBS, named "The Ring of Truth." Now writing a book is mostly a sequestered, reflective process, you at the keyboard with a pile of books and notes; you think back on the past in a quiet familiar room, or sometimes open a conversation with one or another collaborator. Television is conceived in that way, too, on paper, but its birth is quite different: urgent, participant, the events unfold in strange places within sight of wonders.

One summer evening we came to the lagoon of Venice just as the sun set. Our freight barge bore us down the canal, as we sat surrounded by the dozens of cases that held the fragile tools of filmmakers. The waters of the Grand Canal reflected the deepening hues of the sky while the ornamented façades one after another were pierced by glitter. Our treasures of well-wrought lenses from the Orient and fine mechanisms

from Paris seemed in no way strange; travelers had always brought curious wares to this proud and secretive city.

Later I found myself at the eyepiece of a telescope that came to us from Galileo's hand across three centuries. It was too fragile to touch; if today we had one of the caravels of Columbus, certainly no one would sail it off to sea. Later still I sat in a high-performance helicopter as it rose to bank suddenly over bison grazing quietly in the thin air of the Lamar valley. We hovered close to the steep layered cliffs of Specimen Ridge, the more visible since the left side wall of the transparent cabin had been removed to open a still better view for the camera, and an exhilarating intimacy with heights for the passenger.

Nor can I forget the pleasure of the crowd under the mountain in the bright sun as they anticipated the swoop of the racers through the curving streets of the little Pyrenees resort where we first caught up with the Tour de France. I will not forget the cold dawn roadside in Nebraska, nor the view of the ships against the blue Straits from the high ridge of the Rock, nor the neat entries inked in Dr. Cannon's hand on the old glass plate streaked by the little gray bands of star spectra.

Those are the joint pleasures that enter any purposeful travel. Travel and some hard work near home are the motif of the second stage of filmmaking, the actual taking of film. In our case we spent about three half-years on "Ring of

Truth," very roughly a half-year each in preparation, filming, and after the film was in hand.

The first stage resembles the writing of a collective book manuscript. After much research in the library, as well as face to face or over the phone taking counsel with experts, a working script is written and discussed, back and forth in many drafts, for each of the six films. Each of three experienced film producers, each with an associate and an assistant, became responsible for two of the films. The sky castles were transferred to secure foundations through reconnaissance trips the producers took to the suggested locations, trips that demand a subtle mixture of diplomacy and skilled logistics. We were then a group of about a dozen people, the three production teams, writers, researcher, and the indispensable supervision and management of the office and the expensive production budget to come.

Now came filmmaking itself in front of a camera. The work day—or the night—on film location is centered tightly on a six-foot axis. At one end of the axis is the camera lens; at the other end, one or two persons talking before the camera. Each end of the axis is the cynosure of one team of three professionals. We therefore traveled as a group of eight or ten or more; nearly always a friend or a local guide or two have come along.

The end where action takes place is the responsibility of the production team. The producer-director is responsible overall, but in active filming works to judge and direct meaning, gesture, phrase, background: serious exposition helped by all the modes of theater. The associate producer is attentive to continuity; since the event filmed will usually be repeated from a different angle or in a different expressive tone, there is a need to make sure that no visual changes inhibit stitching portions of several different records together into a whole. Hands should stay on the tabletop for one sentence, time after time. The third person, an assistant, does everything else, hair-combing to lunch, and above all makes sure the props are on hand and the parking permit ready to show an inquisitive policeman. The company found meals a joy of leisure, sharing, and the extension of experience.

The production teams stayed together throughout all three phases of the filmmaking. The camera crews are concerned mainly with the taking of film itself. They are assembled anew at each location, working as professionals under contract for a specified number of days. The cameraman, the cinematographer, is the head of the trio focused on the lens end of the axis. His tireless and discerning eye is always at the camera viewfinder from whatever perch allows the right line of sight. He is responsible for ensuring light, shade, detail, and the field of view in width and depth, as the producer may direct. The assistant tends the delicate camera, to feed it magazines of film and new batteries as its appetites demand. The assistant has wide duties, from blowing dust-free gas against a dusty

lens to the setting up of many diverse lights whenever sunlight is not the source. Even the sun may be aided by a diffuse white screen he holds to fill in the blacker shadows. The third member of this team is the sound recordist. The recording microphones pick up speech and ambient sound; the recordist alone hears through headphones just what the microphones heard, and approval for sound is sought before any shot can be accepted. Was that barking dog or distant plane bothersome?

The hour of film viewed on each TV broadcast is a careful mosaic pieced out from all that the camera saw in field or shore, in studio, home, or lab. In our practice an hour of broadcast image is chosen from fourteen to sixteen hours of color film, about six miles of tiny frames 16 millimeters wide, twenty-four frames per second, and forty frames to the foot. That film is complemented by the sound track, transferred from the narrow audio tapes made on the spot to sound on reels of matching 16-millimeter film, cued to the image. Sound and image can be run at the right speed under frame-by-frame control through the editing machine, where it is viewed on a small projection screen and heard on a modest loudspeaker. (The management of many copies, some archived, some roughly processed for immediate work, is beyond our level of detail.)

The film is made up of shots, each one an unbroken roll of the camera, from its start as the striped clapstick signals to eye and ear until motion is stopped by the cameraman. Shots vary widely, but they average perhaps twenty to forty per hour of filming. The editor under the producer's guidance seeks to piece together selections from the hours of film for meaning, drama, and visual qualities. The camera field of view should change, but not too much. The story flows; its movement may be smoothed, but ought not to be made uneventful. This is both technique and art, close to the most active editing of prose. For this phase each production team added a professional film editor.

An hour of finished film is made up of some 300 or 400 shorter selections cut from the 600 shots brought in by the camera, only one-fifteenth of the length of all the film made. A chapter of this book, the prose map of an hour of finished film, has about 600 or 700 sentences. A cut, then, is something like two or three adjoined sentences: it is the length of a fragment of real time that flows past in the film as it did in life. All the longer flow of film, although it may be a mirror of what happened, need not be.

There are more elements to a film. Just as an illustrated book is made of two separate streams, images and text, brought together by the designer in layout, so our films make use of the ingenuity of animation, including the titles and more elaborate video additions to the live camera image. Usually music is added to the soundtrack as well. Those elements flow separately, far apart—our animator was in Long Island—until united in the final film.

In the end we had all together made the films.

The process is comparable to the writing and editing, the procurement and layout of images for a book. Those many technical steps that intervene between printout and a flood of printed books, or between one master print of a film and its distribution to a hundred TV stations are essential, but do not modify content. We have learned to sit quietly to watch the credits unroll; those names denote the professions and the people who have worked artfully on your behalf.

We now append with gratitude the long and worthy list of our friends and collaborators, from our boss, Executive Producer Michael Ambrosino, whose energy, taste, and human warmth are marked on every shot, to indispensable and expert technicians not even mentioned here whom we have never met. Filmmaking is manifestly a collective enterprise, the art form most characteristic of this century, even if our example offers not drama but only the restrained mode of exposition.

The Ring of Truth, A Production of Public Broadcasting Associates, Inc.
Executive Producer: Michael Ambrosino
Production/Business Manager: Marian White
Producer/Directors: Sanford Low, Ann Peck, Terry Kay Rockefeller
Co-Producers: Sheila Bernard, Melanie Wallace
Associate Producer: Peter Frumkin
Production Assistants/Assistant Editors: Mark Henderson, Janet Jennings, Bonnie Waltch
Researcher: Elizabeth Cavicchi
Office Manager: Ellen Murphy Bailey
Intern: Parthiv Shah
Editors: Michel Chalufour, Eric Handley, Lawrence Ross
Cinematographers: Jeri Sopanen, Boyd Estus, Jon Else, Dyanna Taylor
Their assistants: Peter Hawkins, Dick Williams, Roger Haydock, Michael Chin
Sound recordists: Michael Penland, John Haptas, Alex Griswold, John Osborne, John Dildine, Curtis Choy
Animator: Al Jarnow
Music: Guy Van Duser
Funders: Major funding for "The Ring of Truth" was provided by Polaroid Corporation, on the occasion of its fiftieth anniversary. The rest came from the Corporation for Public Broadcasting, Public Television Stations, the National Science Foundation, the Carnegie Corporation of New York, and the Arthur Vining Davis Foundations. We thank them all.

CHRONOLOGY
by Elizabeth Cavicchi

CHAPTER 1: LOOKING

1425   Filippo Brunelleschi—first drawing in perspective

1610   Galileo Galilei—discovers the moon's terrain, new stars, Jupiter's moons, phases of Venus with telescope

1687   Isaac Newton—*Principia:* laws of motion and gravity understood

1800-1   William Herschel—discovers the infrared; J. W. Ritter—discovers the ultraviolet

1887   R. J. Petri—new culture dish for growing bacteria

1887   H. R. Hertz—experimental production of electromagnetic waves

1895   W. C. Roentgen—discovery of X-rays

1912   V. F. Hess—cosmic radiation found in balloon flights

1935   K. G. Jansky—radio waves from space discovered

1969   W. J. Cocke, M. J. Disney, D. J. Taylor—first optical pulsar is discovered in the Crab Nebula

1971   B. Julesz—human visual system senses depth from random dot diagrams printed on flat paper

CHAPTER 2: CHANGE

1789   Antoine Lavoisier—law of mass conservation; quantitative chemistry experiments show that respiration and combustion take up oxygen

1824   Sadi Carnot—heat is produced in every activity and reaction

1842-3   J. R. Mayer, J. P. Joule—theory and experimental verification of the equivalence between mechanical energy and heat

1847   H. L. F. von Helmholtz—conservation of energy proposed

1881   J. K. Starley, J. B. Dunlop—production of "Rover" safety bike with direct steering, diamond frame, two equal-sized wheels, pneumatic tires

1899   W. O. Atwater, F. G. Benedict—energy is conserved by human body

1907   Albert Einstein—$E = mc^2$ formally derived

1933   K. T. Bainbridge—equivalence of mass and energy experimentally verified

1942   Manhattan Project—produces first release of the atom's stored energy in a large-scale reaction

1977   P. MacCready (B. Allen, pilot)—human-powered flight across English Channel

CHAPTER 3: MAPPING

240 B.C.   Eratosthenes—estimate of earth's circumference based on angular height of sun and distance between two cities

c. 1100   Shen Kua, China—compass bearings used in making maps

1507   Martin Waldseemüller—early map showing New World discoveries

1615   Willibrord Snel—maps made by sighting triangles between prominent locations

1696   J.-D. Cassini and French Academy—first scientific world map

1761   John Harrison—first accurate portable clock (chronometer) able to find longitude at sea

1784   C. F. Cassini—first map of an entire nation based on triangulation surveying

1920s   V. Kauffman, T. Abrams, S. M. Fairchild, others—photographs taken from airplanes are used to make maps

1924   L. Mintrop—seismic mapping

1963   H. H. Schmid (U.S. Army)—triangulation mapping using artificial satellites

1967   NASA Lunar Orbiters, ATS-3 (color)—first photos of the whole earth seen from space

CHAPTER 4: CLUES

70 million years ago (mya)—dinosaur extinction

55 mya—Yellowstone eruptions cover the forests of Specimen Ridge

5 mya—Mediterranean sea dries up

1.5 mya—first proto-humans in Africa

1849   M. J. Usiglio—chemical analysis of residue from dried ocean water

1878   W. H. Holmes—fossil forest in Yellowstone described

1950 +   U.S., U.S.S.R., Venezuela—exploratory offshore drilling begins

1970   K. Hsü, W. Ryan, M. Cita and others—*Glomar Challenger* discovers Mediterranean Sea was once dry

1977   B. Heezen, M. Tharp—panorama map of world ocean floor completed

1980   Mount Saint Helens erupts

CHAPTER 5: ATOMS

4000 B.C.   Egypt—gold, silver, lead, copper, glass worked

400 B.C.   Democritus—first idea of atoms

1807   John Dalton—matter made of different atoms, all like atoms have same weight, atoms combine in whole ratios

1814   Joseph von Fraunhofer—first spectroscope, lines in solar spectrum resolved

1869   D. I. Mendeleyev, J. L. Meyer—periodic table of the elements

1898   Marie and Pierre Curie—first purification of radioactive elements

1908   C. G. Barkla—Characteristic X-ray spectra of atoms discovered

1911   Ernest Rutherford—nucleus of atom discovered

1913   Niels Bohr—early quantum model of atom

1913   J. Franck and G. Hertz—experimental evidence for Bohr theory

1919   Ernest Rutherford—first humanly induced transmutation of elements

1925-6   W. Heisenberg, E. Schroedinger—quantum mechanics developed

1980   H. Dehmelt, M. Hohenstatt, W. Neuhauser, P. Toschek—single atoms trapped and held

CHAPTER 6: ASSURANCE AND DOUBT

1687   Isaac Newton—laws of motion and gravity

1755   Immanuel Kant—the spiral patches of light seen through telescopes are galaxies of stars

1827   Félix Savary—first quantitative application of Newtonian gravitation beyond our solar system to double stars

1842   C. J. Doppler—theory that relative speed can be found from frequency of transmitted wave signals

1859   G. R. Kirchhoff—each element has a unique spectrum

1876   W. Huggins—first practical celestial photography

1924   Cecilia Payne—abundances of elements in all stars are the same

1926   E. Hubble—estimate of mass density of the universe (based on optically visible galaxies)

1958   J. H. Oort, F. J. Kerr, G. Westerhout—Milky Way understood as a spiral galaxy from hydrogen radio line

1978   Vera Rubin, Kent Ford, *et al.*—rotation speeds of galaxies too high for visible stars

MORE TO READ

by Elizabeth Cavicchi

Video cassettes of the televised programs on which this book is based can be obtained from:

PBS Video
1320 Braddock Place
Alexandria, VA 22314
800-344-3337
703-739-5380

CHAPTER 1: LOOKING

General Readings:

Album of Science: From Leonardo to Lavoisier, 1450–1800, I. Bernard Cohen, Charles Scribner's Sons, 1980. Inventions and images that revolutionized science are lavishly depicted in woodcuts and engravings from the original sources.

Discoveries and Opinions of Galileo, Stillman Drake, trans., Doubleday, 1957. Galileo's amazing telescopic discoveries, reported in his own words, with historical background.

How to Make a Telescope, Jean Texereau, Allen Strickler, trans., Willmann-Bell, Richmond, Va., 1984. A prominent optician details how to make a telescope, step by step, from grinding the mirror to mounting the tube.

The Mind's Eye, readings from *Scientific American,* Jeremy Wolfe, intr., W. H. Freeman, 1986. A dozen articles on the way the eye sees, how the mind interprets what we see, and optical illusions that show us the workings of eye and mind.

The Birth of a New Physics, I. Bernard Cohen, W. W. Norton, 1985. Arguments that overthrew the ancient belief that the sun orbits the earth.

The Crime of Galileo, Giorgio de Santillana, University of Chicago Press, 1965. The background behind Galileo's trial by the Inquisition.

Perception, Irvin Rock, Scientific American Books, 1984. The visual cues we use to see motion, the third dimension, and upright objects, even though our eyes do not see the world that way.

Powers of Ten: About the Relative Size of Things in the Universe, Philip Morrison and Phylis Morrison and the Office of Charles and Ray Eames, Scientific American Library, 1982. A journey that begins with a picnic and advances to worlds of increasingly large and small dimensions.

Scientific Instruments, Harriet Wynter and Anthony Turner, Charles Scribner's Sons, 1975. The hand-crafted instruments of common use in pre-twentieth-century science, described in text and pictures.

A Short History of Scientific Ideas to 1900, Charles Singer, Clarendon Press, 1959. The ideas of science through the ages are summarized.

More Specialized:

Invention of the Telescope, A. von Helden, American Philosophical Society, 1977. Galileo's telescope and its Dutch predecessors; with the documents.

Galileo at Work; His Scientific Biography, Stillman Drake, University of Chicago Press, 1978. A full study by the most understanding of Galileo scholars.

The Senses, H. B. Barlow and J. D. Mollen, Cambridge University Press, 1982. The functions of our anatomy that reveal the world to us.

"The Space Telescope," John N. Bahcall and Lyman Spitzer, Jr., *Scientific American,* July 1982, p. 40. Plans for an orbiting telescope that will be free of the atmosphere. See *Sky and Telescope,* February 1987, p. 146 for a launch update.

CHAPTER 2: CHANGE

General Readings:

Bicycling Science, Frank R. Whitt and David Gordon Wilson, MIT Press, 1977. The human engine and the mechanics of the vehicles it powers are described and assessed for performance.

Breakaway: On the Road with the Tour de France, Samuel Abt, Random House, 1985. The strategies, trials, and suspense of the 1984 Tour that culminated in an upset win.

Introductory Nutrition, 5th edition, Helen A. Guthrie, C. V. Mosby, 1983. Nutritional requirements, their role in our metabolism, and the deficiency diseases. The caloric content of familiar foods is tabulated.

PSSC Physics, 6th edition, D. C. Heath, 1986. A classic high school text exploring motion, conservation of energy, light, electricity, and the atom; supplemented with problems.

"The Aerodynamics of Human-Powered Land Vehicles," Albert C. Gross, Chester R. Kyle, and Douglas J. Malewicki, *Scientific American,* December 1983, p. 142. Streamlining can allow an athlete to exceed substantially the record speeds achieved on conventional racing bikes.

Space and Time in Special Relativity, N. David Mermin, McGraw-Hill, 1968. The startling results of Einstein's relativity, including the equivalence of mass and energy, are explained with helpful diagrams.

More Specialized:

"Subtle is the Lord...": The Science and the Life of Albert Einstein, Abraham Pais, Clarendon Press, 1982. A richly informative biography, difficult in places.

"Energy and Exercise," Henry A. Bent, *Journal of Chemical Education,* vol. 55 (1978), pp. 456, 526, 586, 659, 726, 796. Humorous articles posing problems about how the body expends energy.

Cycling Physiology for the Serious Cyclist, Irvin E. Faria, Charles C Thomas, Springfield, Ill., 1978. Breathing, food and fluid intake, heat loss, and energy consumption during cycling are analyzed for coach and athlete, from source references.

CHAPTER 3: MAPPING

General Readings:

The Mapmakers, John Noble Wilford, Alfred A. Knopf, 1981. Landmark episodes convey the daring, persistence, and ingenuity of the people who pioneered cartography from Eratosthenes to the Mariner mission to Mars.

Mapping, David Greenhood, Ralph Graeter, illus.,University of Chicago Press, 1964. Practical exercises teach the reader to read map coordinates, contours, and projections and to use surveying techniques.

Road Atlas, Rand McNally, 1986. The biggest bargain in cartography.

Jefferson's Monticello, William Howard Adams, Abbeville Press, 1983. The architectural and personal ideals which guided the design of Thomas Jefferson's renowned home. Magnificently illustrated from his drawings and color photographs of the grounds at Monticello.

The Life of Benjamin Banneker, Silvio A. Bedini, Charles Scribner's Sons, 1972. Biography of the self-taught black astronomer who assisted Ellicott in the surveying of Washington, D. C., supplemented with documents from the time. See also Bedini's *Tinkers and Thinkers*; American innovations in scientific instruments.

Sky and Telescope, Sky Publishing Corp., 49 Bay State Rd., Cambridge, Mass., 02239-1290. Lucid articles on current astronomy, with monthly charts of star and planet positions.

The Astronomical Companion, Guy Ottewell, Department of Physics, Furman University, Greenville, S.C. 29613, 1979. The regularities of the heavens—seasons, star time, phases, eclipses, our position among the stars—are told in large readable diagrams.

Jean-Dominique Cassini and His World Map of 1696, Lloyd A. Brown, University of Michigan Press, 1941. The room-sized map made possible by Cassini's improved method for finding longitude.

Visual Display of Quantitative Information, Edward R. Tufte, Graphics Press, Cheshire, Conn., 1983. A style manual for graphics: clear and effective ways to present numbers in pictures.

More Specialized:

"Who Discovered Mount Everest?" Parke A. Dickey, *Eos*, vol. 66, no. 41 (October 8, 1985), pp. 698–700. The world's highest mountain was first identified by surveyors working on the distant plains of India and clerks who constructed a map from their data.

Elements of Cartography, 5th ed., A. R. Robinson, R. D. Sale, J. L. Morrison, P. C. Muehrcke, John Wiley & Sons, 1984. History and practice of mapmaking.

Scientific Instruments of the Seventeenth and Eighteenth Centuries, Maurice Daumas, B. T. Batsford, trans., Praeger, 1972. The craft and materials that provided new instruments for science.

"Global Positioning System: Refined Processing for Better Accuracy," Javad Ashjaee, *Sea Technology*, March 1986, p. 22. How signals received from orbiting satellites can determine a position on the earth. See the October 1983 issue of *Proceedings of the IEEE* for a more extensive discussion.

CHAPTER 4: CLUES

General Readings:

The Mediterranean Was a Desert: A Voyage of the "Glomar Challenger," Kenneth J. Hsü, Princeton University Press, 1983. An animated recollection of the shipboard debate and hard labor which climaxed in an unexpected discovery.

Geology Illustrated, John S. Skelton, W. H. Freeman, 1966. Aerial views produce evidence of the effects of wind, water, ice, and volcanism on the terrain.

Landprints: On the Magnificent American Landscape, Walter Sullivan, Times Books, 1984. The look of the land, seen from above in photographs, and explained from the geological processes active in shaping it.

Roadside Geology of the Yellowstone County, William J. Fritz, Roadside Geology Series, Mountain Press Publishing Company, Missoula, Mont., 1985. An overview on volcanos, glaciers, and the Yellowstone fossil forest is preface to more than a dozen guided automobile tours. Volumes in this series describe geological formations viewable from roadside in other states.

The Voyage of the "Challenger," Eric Linklater, Doubleday, 1972. The adventurous travels of the nineteenth-century *Challenger* through the world's oceans, splendidly illustrated from works of the time.

Minerals, Lands, and Geology for the Common Defence and General Welfare, vol. 1, *Before 1879*, Mary C. Rabbitt, United States Government Printing Office, 1979. The geologists' explorations of the rugged and mineral-rich territories of the Wild West and the resulting politics back East.

More Specialized:

"The Petrified Forests of Yellowstone Park," Erling Dorf, *Scientific American*, April 1964, p. 107. The fossilized remains from two dozen stacked layers of ancient forests are dated and identified.

Report of Scientific Results of the HMS "Challenger," C.W. Thomson and J. Murray, Johnson Reprint Co., 1895/1965. Thirty volumes containing the wealth of new knowledge gained from a sailing expedition through the oceans.

"History of the Mediterranean salinity crisis," K. J. Hsü et al., *Nature*, vol. 267 (1977), p. 399. Core samples from beneath the seafloor provide clues that the Mediterranean once dried up. See also *Scientific American*, December 1972, p. 26.

Initial Reports of the Deep Sea Drilling Project, University of California, Scripps Institute of Oceanography, in 80+ volumes, 1969–1983. The entire results from *Glomar Challenger*'s worldwide voyages that transformed geology. See vol. 13, part 2, and vol. 42 for the Mediterranean Sea story.

CHAPTER 5: ATOMS

General Readings:

Descriptive Chemistry, Donald A. McQuarrie and Peter A. Rock, W. H. Freeman, 1985. Properties of the chemical elements, alone and in compounds, are displayed through examples, questions, and vivid photography.

A Search for Structure: Selected Essays on Science, Art, and History, Cyril Stanley Smith, MIT Press, 1981. A metallurgist and historian reflects upon a lifetime of inquiry into the interwoven history of materials, craftsmanship, and the arts.

Crystals and Crystal Growing, Alan Holden and Phylis Morrison, MIT Press, 1982. The structures built by atom-sized blocks are explored, with tested recipes for growing them from common ingredients.

Knowledge and Wonder: The Natural World as Man Knows It, Victor F. Weisskopf, MIT Press, 1979. The big picture science has drawn of the natural world, from the atom's quantized jumps to the origin of life, made clear and concrete.

More Specialized:

Discovery of the Elements, Mary Elvira Weeks and Henry M. Leicester, *Journal of Chemical Education,* Easton, Pa., 1968. A chronicle with anecdotes about the people who first purified each of the elements.

The Cambridge Guide to the Material World, Rodney Cotterill, Cambridge University Press, 1985. How the manifold variety of the world around us is a direct result of microscopic behaviors and orderings of atoms.

Niels Bohr: A Centenary Volume, A. P. French and P. J. Kennedy, Harvard University Press, 1985. The creative work, ideals, and life of the man who cracked the atom's quantum code, remembered by his students.

Nobel Lectures in Physics, 1922–1941, Nobel Foundation, Elsevier, 1965. The acceptance speeches of Franck and Hertz for the 1925 prize are included in this volume.

"Cooling and Trapping Atoms," William D. Phillips, Harold J. Metcalf, *Scientific American,* March 1987, p. 50. Intricate traps slow and catch atoms to allow precise study of their behavior.

Chapter 6: Assurance and Doubt

General Readings:

Cecilia Payne-Gaposchkin: An Autobiography and Other Recollections, Katherine Haramundanis, ed., Cambridge University Press, 1984. Cecilia Payne's eloquent story of her childhood desire to become a scientist and of the difficult but productive years at Harvard.

The New Astronomy, Nigel Henbest and Michael Marten, Cambridge University Press, 1983. Features of the universe newly revealed by light in wavelengths from the long radio to the short X-rays are described and handsomely illustrated.

Astronomy: The Cosmic Journey, William K. Hartmann, Wadsworth, 1982. An introductory text tells of the worlds beyond ours as we see them: the solar system, the stars, the galaxies.

The Lighter Side of Gravity, Jayant V. Narlikar, W. H. Freeman, 1982. The force of gravity causes both familiar and unusual effects.

"Dark Matter in Galaxies," Vera Rubin, *Scientific American,* June 1983, p. 96. Measurements taken far out on the long arms of galaxies show that the speeds of rotation there are much higher than expected. Perhaps unseen matter in those regions boosts the speeds.

More Specialized:

A Source Book in Astronomy and Astrophysics, 1900–75, Kenneth R. Lang and Owen Gingerich, eds., Harvard University Press, 1979. Modern results that transformed astronomy are charted in excerpts from the discoverers' papers. Payne's work and its later confirmations are included.

The Analysis of Starlight: One Hundred and Fifty Years of Astronomical Spectroscopy, J.B. Hearnshaw, Cambridge University Press, 1986. The history of dissecting stellar spectra to reveal the character of stars, from before the photographic plate to the theoretical techniques of modern physics.

Stellar Atmospheres, Cecilia H. Payne, Harvard Observatory, 1925. The spectra of the stars are unraveled to yield clues of their temperature and composition in this extraordinary thesis.

"Quantum Physics and the Stars I and II," David H. DeVorkin and Ralph Kenat, *Journal for the History of Astronomy,* vol. 14 (1983), pp. 102 and 180. The historical setting for the astrophysical problems faced by Payne and Russell.

Histories

There are several encyclopedic historical works that make good browsing and authoritatively supply oceans of fact. They were much used in making this book. Students will find them helpful; they are widely held in libraries, expensive though rewarding for bookish homes.

Joseph Needham and others, *Science and Civilization in China:* Cambridge, 1954–   . A dozen thick books and still coming; the seven "volumes" may have several parts, each a full-sized book. They are unique in tracing the history of science and technology in China, but the valuable surprise is that for many topics they include by way of comparison a concise but rich survey of events and ideas in Europe and elsewhere.

Charles Singer, E. J. Holmyard, and A. R. Hall, *A History of Technology,* chronologically arranged in five big volumes, Oxford, 1954. No longer the last word, the articles are so expert and comprehensive, and so fascinatingly illustrated, that the work is still a primary resource for the period from flint-making up to about 1900.

Concise Dictionary of Scientific Biography: New York, 1981. This heavy volume has thumbnail accounts of the lives of about 5,000 contributors to science and technology from all periods and places, without illustrations. It is a stand-in for its sixteen-volume parent under editor C. C. Gillispie.

INDEX

by S.W. Cohen and Associates